TRANSACTIONS OF THE JAPANESE SOCIETY FOR NON-DESTRUCTIVE INSPECTION

Volume 4

Transactions of the Japanese Society for Non-Destructive Inspection

Each volume contains original contributions in English and translations of papers published in Japanese in *Hihakaikensa* (the journal of the Japanese Society for Non-Destructive Inspection).

Transactions of the JSNDI, Volume 1
Edited by Teruo Kishi
Published by the JSNDI

Transactions of the JSNDI, Volume 2
Edited by Hiroshi Kato

Transactions of the JSNDI, Volume 3
Edited by Hiroshi Kato

Transactions of the JSNDI, Volume 4
Edited by Ichirou Yamaguchi

TRANSACTIONS OF THE JSNDI

Volume 4

Translations from the Japanese
and original contributions in English

Published on behalf of the Japanese Society for Non-Destructive Inspection

GORDON AND BREACH SCIENCE PUBLISHERS

Philadelphia • Reading • Paris • Montreux • Tokyo • Melbourne

Japanese Society for Non-Destructive Inspection

5-4-5 Asakusabashi
Taito-ku, Tokyo 111
Japan

Gordon and Breach Science Publishers

5301 Tacony Street, Drawer 330
Philadelphia, Pennsylvania 19137
United States of America

Post Office Box 161
1820 Montreux 2
Switzerland

Post Office Box 90
Reading, Berkshire RG1 8JL
United Kingdom

3-14-9, Okubo,
Shinjuku-ku, Tokyo 169
Japan

58, rue Lhomond
75005 Paris
France

Private Bag 8
Camberwell, Victoria 3124
Australia

Library of Congress Cataloging-in-Publication Data

Hi-hakai kensa. English.
   Transactions of JSNDI translated from Japanese
      p. cm.
   Title on added t.p.: Transactions of the Japanese Society for Non
-Destructive Inspection.
   "Each v. contains English translation of papers originally
published in Japanese in Hi-hakai kensa."--verso of v. 2, t.p.
   Includes index.
   ISBN 2-88124-793-8 (v. 4)
   1. Non-destructive testing. I. Nihon Hi Hakai Kensa Kyōkai.
II. Title.   III. Title: Transactions of the Japanese Society for Non
-Destructive Inspection.
TA417.2.H5123   1992
620.1'127-dc20   90-13821
                                                                  CIP

Contents

TEMPERATURE DEPENDENCE AND FRACTURE STRENGTH OF POLYMERS IN IMPACT TESTING*

AKIRA SHIMAMOTO, FUMIO NOGATA† and SUSUMU TAKAHASHI‡

Associate Professor, Department of Mechanical Engineering, Saitama Institute of Technology, 1690 Fusaiji, Okabemachi, Osato-gun, Saitama, 369-02 Japan

The Charpy impact test has been widely applied to examine the strength of engineering plastic. However, there are some differences among the methods of this test as it is standardized in each nation in aspects of notch shape, test specimen dimensions, and impact velocity. The present study investigated fracture strength of certain engineering plastics for a range of temperatures varying from $-196°C$ to $20°C$; fracture strength was calculated from impact energy, and fractography was also done for the transition temperature region of the plastic material. Unexpectedly, polycarbonate material was found durable up to $-100°C$ and no sudden critical region was observed as with metal material, meaning that the low temperature brittle fracture occurred at $-140°C$ or below. In cellulose acetate, cellulose propionate and polyimide, however, the sudden critical region was observed, and a remarkably temperature-dependent low temperature brittle fracture took place.

KEY WORDS: Charpy Testing, Engineering Plastics, Material Testing, Fractography.

1. INTRODUCTION

In recent years polymers have been replacing metals in industrial and structural materials, creating new substances of higher elastic modulus and strength; the term *engineering plastics* has been used for these since the 1950s. When polymers are used as industrial materials, however, arguments often arise relative to the impact strength of their mechanical properties. The reasons for the different opinions are that there are many types of tests conducted using different shapes of specimen with every applicable standard; this means that: (i) a comparative evaluation of test results is impossible, (ii) impact values obtained are not correlated with other physical characteristics, and (iii) impact strength values in terms of energy cannot be used for design unless they are converted into stress. The authors[1-4] earlier clarified (i) and (ii) above. In this report, a theoretical study is detailed concerning the effect of specimen shape, dimensions and temperature on the Charpy impact values using a theoretical formula. Fracture appearances obtained in the transition temperature range were also examined.

* Originally published in Japanese in Hihakaikensa [J. JSNDI] vol. 37/No. 1 (1988)

† Associate Professor, Department of Mechanical Engineering, Himeji Institute of Technology, 2167 shosha, Himeji, Hyogo, 671-22 Japan

‡ Professor, Department of Mechanical Engineering, Kanto Gakuin University, 4834 Mutuuracho, Kanazawa-ku, Yokohama, 221 Japan

2. THEORETICAL ANALYSIS OF CHARPY IMPACT VALUE

Rupture of a Charpy impact specimen occurs when the maximum stress at the bottom of a notch due to bending[7] caused by a hammer blow reaches the rupture strength (limit stress) σ_c of the material; the absorbed impact energy can be expressed as the elastic strain (deflection) of energy stored in the specimen up to that moment[8]. The elastic waves of the blow have gone back and forth through the specimen several tens of times before rupture occurs[9], and the deflection energy can be calculated as a quasi-static factor of the strength of the material.

Assuming F is the load imposed at the center of a beam specimen of width b, thickness t, modulus of longitudinal elasticity E and Poisson's ratio ν, which is supported at both ends with a distance between supporting points S, and compliance C (when a notch is provided) is K times Co, the strain energy stored in the specimen can be expressed as follows:

$$U_E = \frac{1}{2}F^2c = \frac{1}{2}KF^2c_0 = \frac{S^3}{8Ebt^3}F^3 \qquad (1)$$

From the definition of stress concentration factor α[7], the maximum stress σ max generated at the bottom of the notch, the stress concentration factor of which is α, generated by the load F is:

$$\alpha = \frac{\sigma_{max}}{\sigma_0} \qquad (2)$$

where σ_0 is a reference stress. If it is assumed to be the stress calculated for the minimum cross section at the bottom of notch, then

$$\sigma_0 = \frac{3}{2} \cdot S/b(t-a)^2 \cdot F \qquad (3)$$

where a is the depth of the notch.

From (1), (2) and (3), the following relation formula can be introduced between the deflection energy U_E and the maximum stress σ max at the bottom of the notch.

$$U_E = q \cdot \frac{K}{\alpha^2}\left(1 - \frac{a}{t}\right)^4 \cdot \frac{btS}{18E} \cdot \sigma_{max}^2 \qquad (4)$$

where, q is a correction factor of shearing which is $q = 1 + 2(1.2 + 1.1\nu)\,t^2/S^2$. The cross sectional area A at the bottom of the notch is,

$$A = b(t-a) = bt(1 - a/t) \qquad (5)$$

Hence, if rupture is assumed to occur at the moment of σ max $= \sigma_c$, Charpy impact value a_{kc} is,

$$a_{kc} = \frac{U_E}{A} = q \cdot \frac{K}{\alpha^2}\left(1 + \frac{a}{t}\right)^3 \frac{S}{18E} \cdot \sigma_c^2 \qquad (6)$$

Thus, if the factor α corresponding to the shape of the specimen used, K and the rupture strength σ_c of test material are known, the Charpy impact value can be calculated.

The increase factor K of compliance C which changes with the dimensions of the notch can be obtained from experimental data. The effect of distance S between supporting points near the notch of a normally used Charpy impact specimen can be expressed as equation (7)[8],

$$\alpha(S) = \alpha_M(1 - 0.3\ t/S), \text{ where } S \geqq 4t \tag{7}$$

in which α_M and $\alpha(S)$ are stress concentration factors for pure bending, and values are obtained as a combination of results of photoelastic tests in pure bending and calculated values based on the finite element method[8], while stress concentration factor α_M is dependent upon thickness of the three point specimen t, the distance between supporting points S, and notch shape, although these latter values cannot be obtained analytically in every case. The increase factor K of compliance C which changes with the dimensions of the notch can also be experimentally obtained.

3. EXPERIMENTAL RESULTS AND DISCUSSION

Once the limit stress σ_c for rupture, stress concentration factor α and the increase factor K of compliance C due to the notch are determined, the Charpy impact value a_{KC} for a specimen of certain configuration can be calculated. Here, using previously reported results[2-4], the rupture strength (limit stress) σ_c was inversely calculated from the Charpy impact value a_{kc}, using equation (6) to learn the effect of changes in dimensions, shape and temperature on this value with notch unchanged, and on the assumption that the limit stress σ_c is constant regardless of specimen shape, for comparative purposes. Results are shown in Figures 1 to 4.

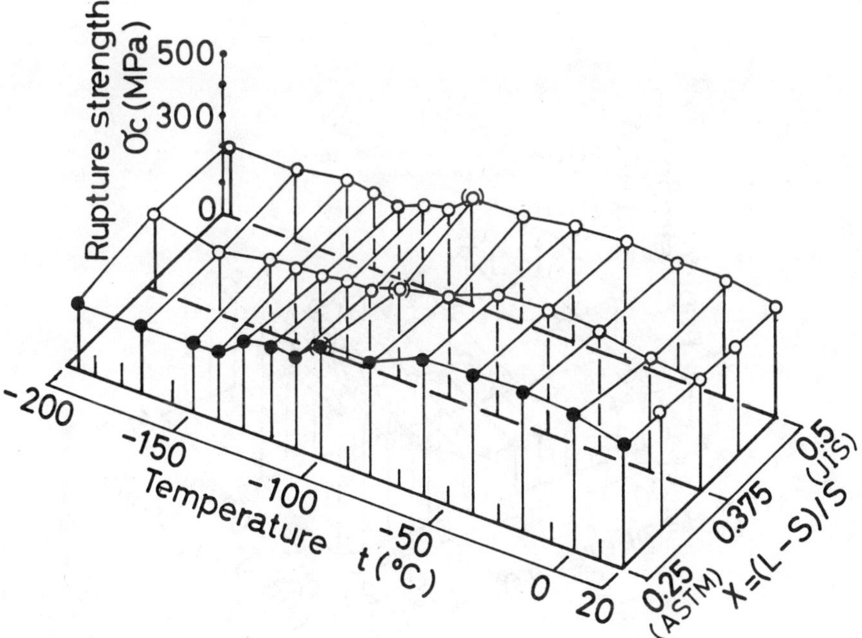

Figure 1 Rupture strength σ_c obtained by inversely calculating from Charpy impact value on polycarbonate

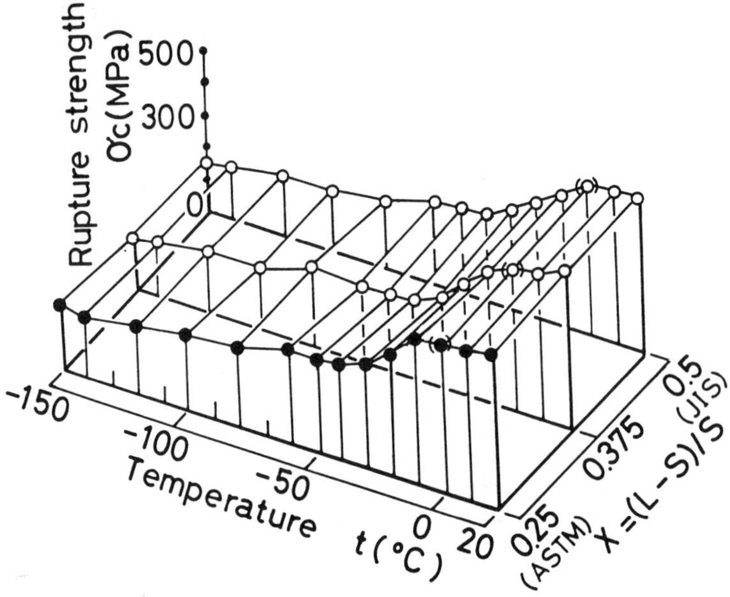

Figure 2   Rupture strength σ_c obtained by inversely calculating from Charpy impact value on cellulose acetate

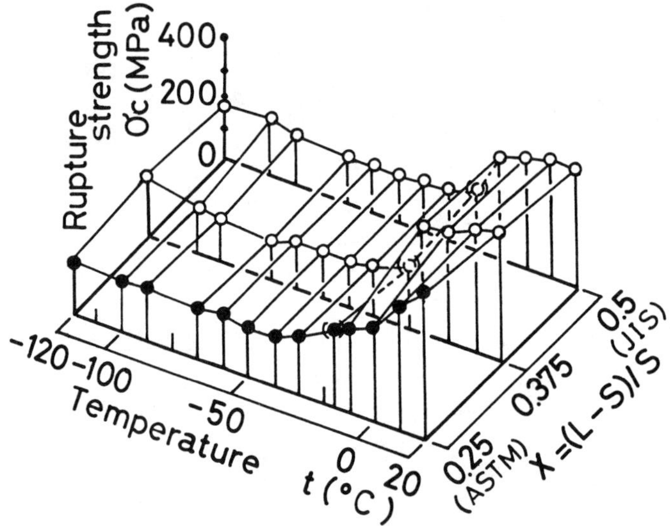

Figure 3   Rupture strength σ_c obtained by inversely calculating from Charpy impact value on cellulose propionate

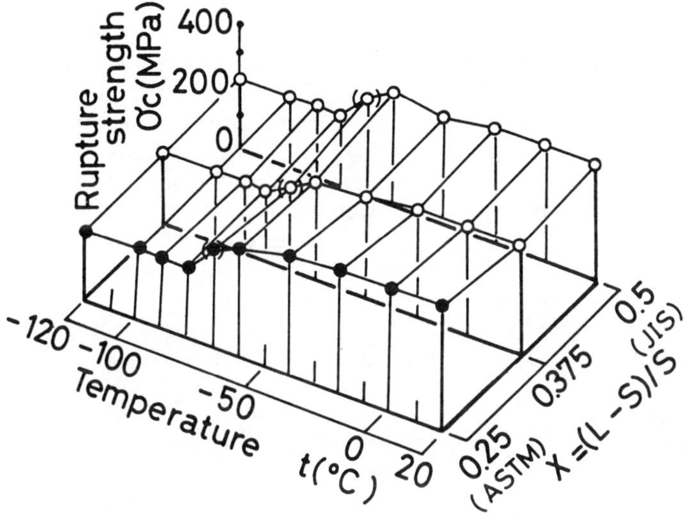

Figure 4 Rupture strength σ_c obtained by inversely calculating from Charpy impact value on polyimide (TI-5000)

3.1. *Effect of specimen shape and distance between supporting points*

As can be seen from Figures 1 to 4, the rupture strength (limit stress) σ_c decreases in proportion to decreasing temperature for specimens of the same material. For a certain temperature range, ($-140°C$ to $-196°C$ for polycarbonate, $-120°C$ to $-150°C$ for cellulose acetate, $-70°C$ to $-120°C$ for cellulose propionate, $-80°C$ to $-196°C$ for polyimide; Figures 1 to 4 respectively), however, the rupture strength σ_c is maintained constant showing no effect of temperature, confirming the same trend as the relation between impact value and temperature.

This seems to be due to complete brittle failure taking place at a given temperature rather than in response to the effect of specimen shape in different standards (e.g. JIS, ASTM), or to the distance S between supporting points.

3.2. *Effect of temperature on impact value*

Considerable possible errors in notch shape during machining, experimental errors including unavoidable unevenness of specimen material and errors in factor α and K values used in the calculation, the results in Figures 1 to 4 indicate that the rupture strength (limit stress) σ_c of each material is nearly constant at each temperature. In polycarbonate, the rupture strength (limit stress) σ_c increases slightly with the lowering of temperature from $20°C$ to $-60°C$; then, as the temperature continues to fall from $-60°C$ to $-140°C$, it gradually decreases. This seems to be due to the size of plastic deformation zone as is true of the impact value a_{kc}—temperature relationship. No abrupt transition zone such as that seen with metallic materials is noticed. Thus, the applicability of polycarbonate down to $-100°C$ has been confirmed. Cellulose acetate shows decrease in rupture strength σ_c in proportion to lowering temperature from 20°C to -120°C, with an especially rapid decrease at around -20°C. Cellulose propionate also decreases in rupture strength σ_c in proportion to a reduction in temperature from 20°C to -70°C, indicating a rapid decrease at around -10°C. In both cases, the size of the plastic deformation zone is small, which seems to be the reason for the above-described trend. In polyimide, the rupture strength σ_c decreases as the temperature falls from 20°C to -80°C, indicating a sudden strength reduction in the vicinity of -70°C, seemingly due to a large plastic deformation zone. In previous reports[2-4] where experiments were conducted on the effect of notch shape and temperature on Charpy impact value, a range was noted for every material where a minimum impact value a_{kc} was obtained, and the fracture appearance of a specimen broken in that range showed an extremely smooth glossy surface or mirror pattern. Moreover, in the case of a slit notch (sharp notch), experimental data showed the fracture behaviour to differ considerably from other cases. This was also reported by Takahashi[11] and Uemura[8].

As to the absolute values of rupture strength σ_c obtained by reverse calculation, their appropriateness was evaluated for those of V notch specified in the JIS standard and in the temperature ranges above and below the transition temperature, respectively, while the total average of values plotted in Figures 1 to 4 can be used in another way.

Indicated in Table 1 is the mean of rupture strength values for each material obtained by averaging these values within the area that is entered by the transition temperature zone, as shown in Figures 1 to 4 for respective materials. The mean deviation ratio for the part of the area corresponding to temperatures lower than the transition temperature zone is seen to be approximately 3%, whereas that for the area corresponding to temperatures higher than this is about 5%. In any event, rupture strength may be regarded as constant within the part of the area applicable. Ratios for different materials of the mean σ_c to respective tensile strength values σ_u obtained by high speed tensioning (100 mm/min.), σ_c/σ_u, are in the range of 2.2–6.9 for temperatures lower than the transition temperature zone, or 3.5–13.6, considerably wider, for temperatures above this zone. In polyimide (TI-5000), in which a brittle failure occurs

Table 1.   Comparison between rupture strength σ_c and tensile strength σ_u

Material	Transition Temperature °C	Mean σ_c (1) value (MPa)	Mean deviation (2) ratio of σ_c (%)	Tensile strength (3) σ_u (MPa)	$\dfrac{\sigma_c}{\sigma_u}$
Polycarbonate	Above -100	346.4	2.2	65.2	5.3
(PC)	Below	230.7	2.9		3.5
Cellulose	Above 0	466	5.6	34.3	13.6
acetate (CA)	Below	236.7	5.8		6.9
Cellulose	Above -15	346	4.1	37.3	9.28
propionate (CP)	Below	163	1.3		4.4
Polyimide (PI)	Above -70	355.8	5.7	103	3.5
(TI-5000)	Below	228.3	0.7		2.2

Note: (1) Mean value of rupture stress values obtained from Charpy impact values.
(2) Average of absolute values of deviation ratio (ratio of individual values to mean value of differences between individual values and mean value).
(3) σ_u obtained by a tension speed of 100 mm/min.

in the transition temperature zone, σ_c/σ_u is about 2. For other materials, even in the temperature range lower than the transition temperature zone, the values of σ_c, inversely calculated by applying the ratio of 3–7, are found considerably larger than σ_u values obtained by static tensile tests. The lower the σ_c/σ_u ratio is, the smaller the elongation of the material at the tensile test will be, resulting in a brittle rupture; polyimide is a typical example. Among those in Table 1, materials having a large σ_c/σ_u ratio, 7 for example, show considerable ductility in a tensile test. The stress condition generated at the bottom of the notch of an impact test specimen at the time of load application is one in which the plastic deformation is strongly restricted due to the notch in addition to simple or multiaxial stress. In general, rupture stress increases considerably in the case of restricted plastic deformation[12]. In a tensile test conducted on a round mild steel bar having a circumferential notch, the smaller the radius at the bottom of the notch is, the more the corresponding stress-strain curve will be shifted towards the brittle fracture side. When a sharp V-notch is applied, the corresponding stress-strain curve becomes similar to that of a cast iron specimen, the rupture stress rising to twice as much or more[13]. These features may be understandable from Table 1 where the different levels of σ_c/σ_u value are seen at different sides of the transition temperature zone.

Results showed that, taking into consideration that the theoretical formula for the Charpy impact value a_{kc} given as equation (6) is based on the rupture strength σ_c

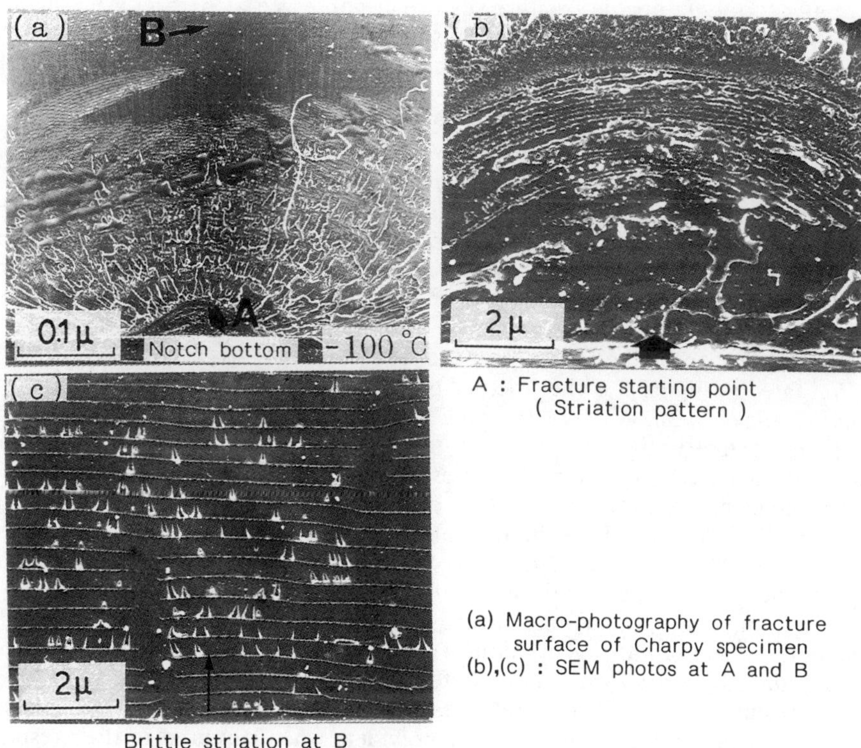

Figure 5   Observed positions and patterns of fracture surface on polycarbonate

calculated using the modulus of longitudinal elasticity E obtained from a static test, the absolute value of rupture strength σ_c obtained by reverse calculation can be considered appropriate. In other words, the theoretical formula for Charpy impact value a_{kc}, equation (6), is in very good agreement with experimental results of low temperature brittle failure.

3.3. *Observation of fracture appearance obtained in transition temperature range*

Figures 5 to 8 are microphotographs of fracture appearance exhibited by different materials ruptured in respective transition temperature ranges.

Figure 5 shows the fracture pattern of polycarbonate (temperature $-100°C$). When examined microscopically, a clear striation pattern is seen in the area of crack initiation A, (this pattern is frequently seen in the area of crack propagation), and a coexisting spider's thread pattern generated by external force at the tip of the crack initiation area. Indicated at the front end B in the direction of crack propagation is a fracture pattern of brittle striation which looks flat and is accompanied with little plastic deformation.

Figure 6 is the fracture pattern of cellulose acetate (temperature $0°C$). Observed at the point of crack initiation A, when examined microscopically, is a river pattern often seen at the early stage of rupture which indicates that the rupture has been initiated and propagated from inside. Elongated parabolic dimples which are normally seen on a ductile fracture surface are observed at Area C. The surface of Area D showing transverse dark bands is somewhat irregular where the dimple pattern is replaced with an intergranular fracture surface which generally results from a low-temperature brittle rupture. Observed at Area E is a granular pattern mixed with congregated flaky spots.

The fracture pattern of cellulose propionate (temperature $-15°C$) is shown in Figure 7.

When microscopically examined, a tear-like pattern is observable at the point of crack initiation A progressing in the direction of crack propagation and a parallel fringe at the front of the notch base. A base-rock pattern indicating an extremely irregular rock-like appearance is seen at Area B which turns predominantly into a mirror pattern of minimized irregularity as it approaches to Area C within the final rupture zone at the end of crack propagation.

Presented in Figure 8 is the fracture pattern of polyimide (temperature $-70°C$). Microscopic examination reveals a ragged tear-like parabolic pattern is indicated at the point of crack initiation A progressing in the direction of fracture propagation; at the front end C in the direction of progress, a Walner line that is normally produced in an extremely brittle material as a result of interference between a propagating crack tip and elastic impacts is seen. Indicated in Area B between the point of crack initiation A and Area C at the front end of fracture propagation are vertically-running ragged and radiative lines with facets of flat surface having shallow holes here and there. At Area D, tongue-like protrusions characterize a fracture mode called "Tongue" where rupture occurs along a particular crystal face called a cleavage plane accompanied by virtually no plastic deformation.

As a result of the fracture surface observations reported above, it was confirmed that the fracture pattern developed by polymers when ruptured in their respective transition temperature range is a mixture of various fracture modes.

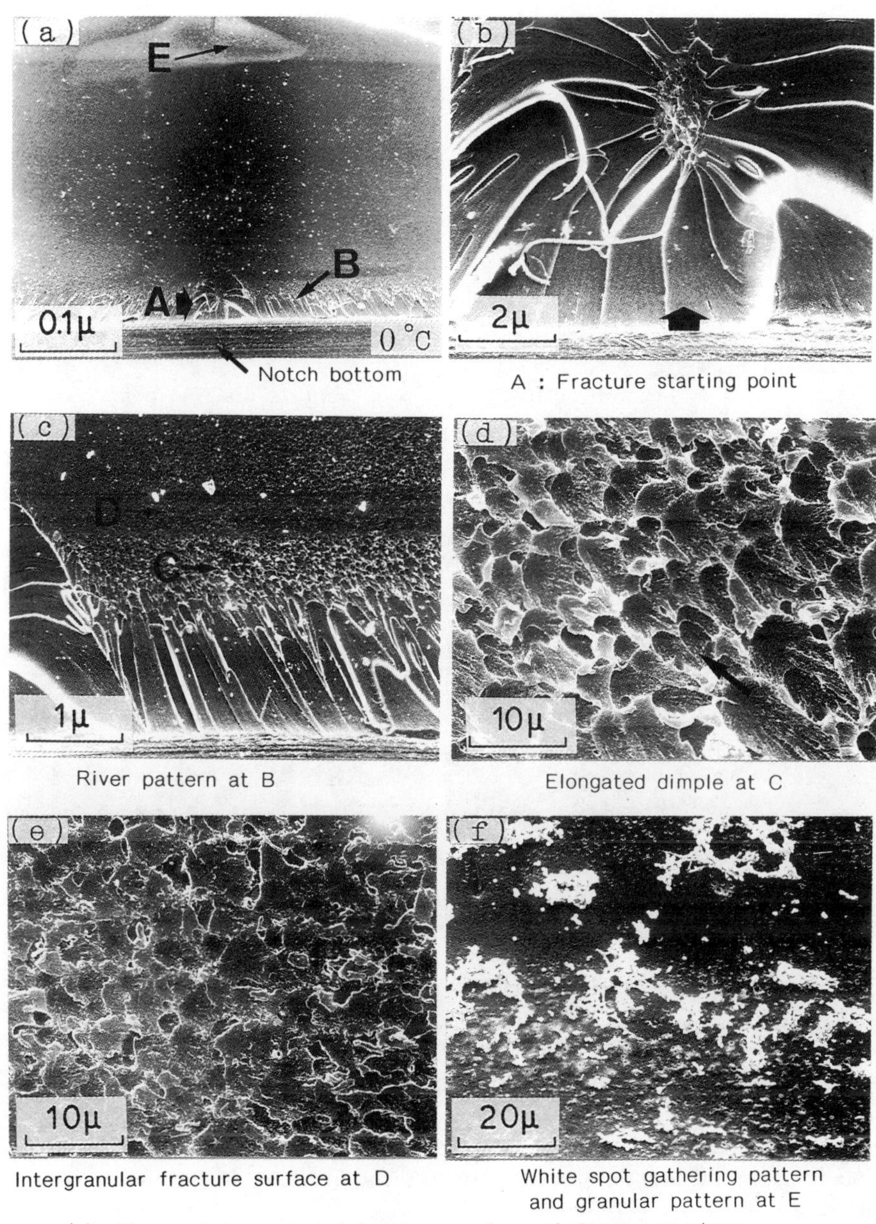

(a) Macro-photography of fracture surface of Charpy specimen

(b), (c), (d), (e), (f) : SEM photos at A, B, C, D and E

Figure 6 Observed positions and patterns of fracture surface on cellulose acetate

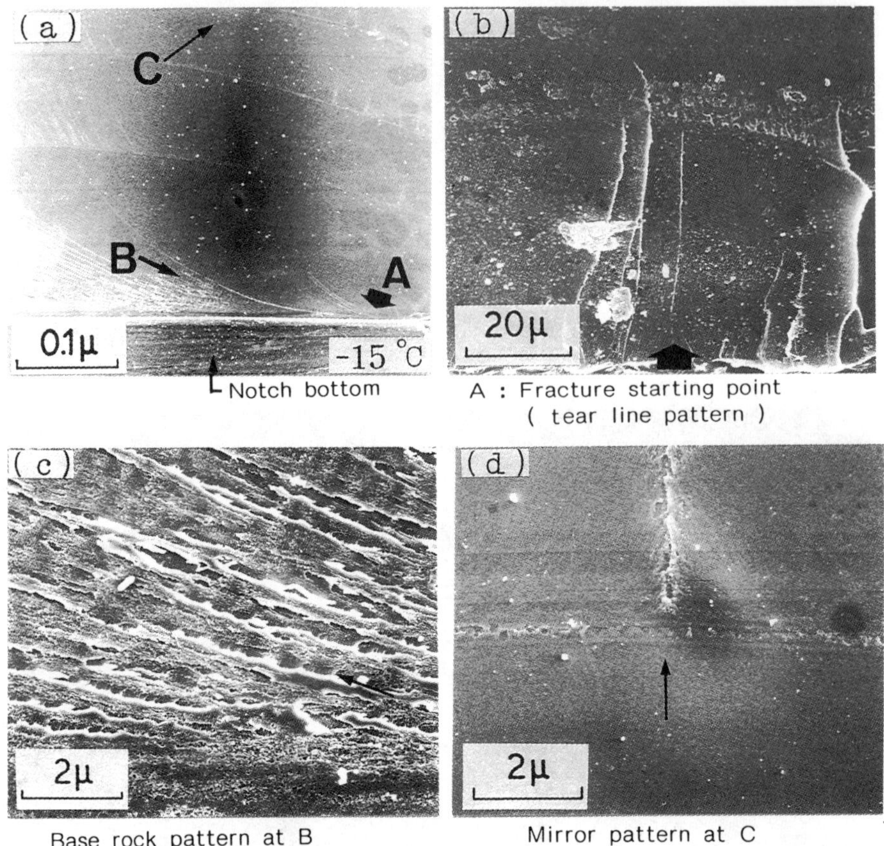

(a) Macro-photography of fracture surface of Charpy specimen
(b), (c), (d) : SEM photos at A, B and C

Figure 7   Observed positions and patterns of fracture surface on cellulose propionate

4. CONCLUSION

Rupture strength (limit stress) σ_c was obtained by reverse calculation from impact value for four different types of polymer, and fracture surfaces examined. The following findings were obtained.

(1) In polycarbonate, no rapid change in rupture strength σ_c was observed when temperature was lowered (20°C to $-60°$C) as is observed in cellulose acetate (20°C to $-120°$C) and cellulose propionate (20°C to $-70°$C).

(2) With the condition of a constant notch shape, the effect of specimen shape of various Standards and temperature changes (low temperature) on both experimental value a_{kc} and rupture strength σ_c was clarified.

(3) Fracture pattern obtained in the transition temperature range for different

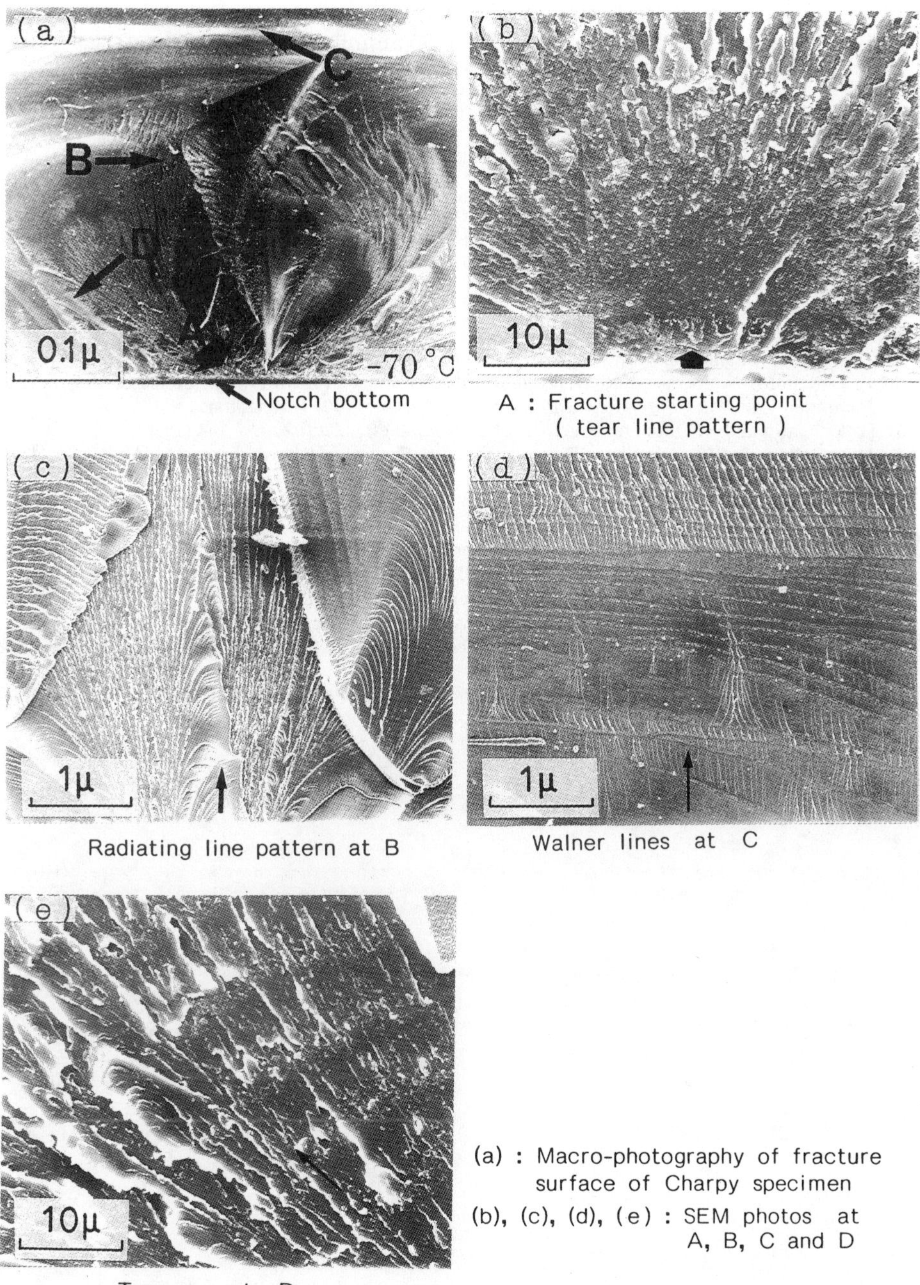

Figure 8 Obsreved positions and patterns of fracture surface on polyimide

polymers was found to be a mixture of multiple fracture modes, and the direction of fracture propagation could be identified by examining the fracture surface.

REFERENCES

1) A. Shimamoto, F. Nogata, S. Takahashi and S. Kumagai: A Consideration on the Charpy Impact Test for Plastics, *J. Jap. Soc. for Non-destructive Inspection*, **34**(5), 341–344 (1985).
2) A. Shimamoto, F. Nogata, S. Takahashi and S. Kumagai: Temperature Dependence of Polymers in Impact Testing, *J. Jap. Soc. Non-destructive Inspection*, **36**(3), 214–222 (1987).
3) A. Shimamoto, S. Takahashi, F. Nogata and S. Kumagai: Temperature Dependence of Polymers in Charpy Impact Testing, *Proceedings of APCS—***86**, 81–86 (1986).
4) A. Shimamoto, F. Nogata and S. Takahashi: Temperature Dependence of Polymers on Impact Testing, *Trans. Jap. Soc. for Non-destructive Inspection* **1**, 218–231 (1988).
5) T. Uchimoto, N. Kawase and H. Tomita: Observation of Fracture Appearance of Polymers (in Japanese), *The 13th Conference of Fractography Section*, **130**, 1–28 (1981).
6) H. Kitagawa and R. Onodera: Fractography (in Japanese), *Baifukan*, 4–144 (1980).
7) J. D. Colton, and G. Herrmann: Dynamic Fracture Process in Beams, *Trans. ASME. Ser. E*, **42**(2), 435–439 (1975).
8) Y. Uemura: Study on Charpy Impact Testing of Hard Plastics, *Trans. Jap. Soc. Mech. Eng.* **46**(407), 815–822 (1980).
9) H. Okamura, T. Takano and S. Mihashi: Experimental Technique of Dynamic Photoelasticity Using Frashlight, (in Japanese), *Trans. JSME*, **38**(310), 1175–1182 (1972).
10) M. Nishida: Stress Concentration (in Japanese), *Morikita Publication*, 15–17 (1967).
11) K. Takahashi: *Wisenschaftlicher Bericht des Instituts für Festkör Permechanik* W 5/1977, 1 (1977).
12) H. W. Hayden, et al.: *The Structure and Properties of Materials*, Vol. III, 137 (1965), John-Wiley & Sons.
13) H. E. Davis, et al.: *The Testing and Inspection of Engineering Materials*, 3rd. ed., 134 (1955). McGraw-Hill.

NONDESTRUCTIVE EVALUATION OF POROSITY IN SINTERED IRON BY MAGNETOMECHANICAL ACOUSTIC EMISSION*

NOBORU SHINKE, YOSHITUGU OHIGASHI, AKIKAZU NAKAGIRI and TAKAAKI YAMANO

Faculty of Engineering, Kansai University, Yamate 3-3-35, Suita, Osaka 564, Japan

The discontinuous motion of magnetic domain walls in ferromagnetic materials during magnetization produces magnetomechanical acoustic emission (MAE). The MAE has characteristics which show a sensitive dependence on metallographical structure of ferromagnetic materials. In this paper, the possibility of nondestructive evaluation of porosity in sintered iron has been mainly investigated by using the MAE. The results obtained are summarized as follows:

(1) The higher porosity in sintered iron increases the MAE intensity, either AE count or amplitude.

(2) Evaluating the effects of porosity with the values normalized by the MAE intensity in bulk material, it is recognized that there is more remarkable porosity effect on mean event energy of the MAE.

(3) The results of measurements suggest an adequate correlation between the porosity and the MAE in sintered iron.

KEY WORDS: Porosity, Acoustic emission, Nondestructive evaluation, Sintered materials, Ferromagnetic materials.

1. INTRODUCTION

As the sintered iron has a unique structure which contains many pores in it, the mechanical and other properties change greatly according to the porosity, size, form and distribution of the pores. Especially, as the porosity has such a fundamental characteristic that the physical and mechanical properties of sintered materials can be estimated, many reports on how the mechanical characteristics depend on the porosity have been made.[1-6] Although the Archimedes method can measure only the average porosity, obtaining local porosity could estimate the condition of sintered iron more accurately.[7-8] It was reported in 1975 that the phenomenon of discontinuous magnetization in the process of magnetizing ferromagnetic materials works to generate magnetomechanical acoustic emission (MAE).[9] After that, it was pointed out that MAE are generated in the process of discontinuous shifts of the 90° magnetic domain wall and rotation of magnetizing vector.[10-11] An interest in MAE is comparatively new, and various experiments have been conduced recently to utilizing MAE as a new method of nondestructive evaluation of ferromagnetic materials.[12-15] In this paper, we will report on the results of a test in which we measured the MAE

* Originally published in Japanese in Hihakaikensa [J. JSNDI] vol. 39/No. 3 (1990)
† Osaka Titanium Co., Ltd., Higashihama 1, Amagasaki, Hyogo 660, Japan.

generated from the sintered iron of several different porosity during magnetization, and from which we have investigated the possibility of nondestructive evaluation of the porosity by the MAE.

2. EXPERIMENTAL APPARATUS AND METHOD

One of the most basic ways to estimate the MAE of sintered irons is to measure AE count in the period of magnetizing frequency, but at the same time it is very important to know the amplitude distribution of the MAE. In this research, we adopted the ring-down counting method for evaluating MAE generated from sintered iron. The automatic analyzer used to measure MAE generated in the process of magnetizing sintered irons is shown in Fig. 1. This apparatus is the same as the one previously reported.[16] We used a PZT transducer with resonance characteristic set at 250 KHz, as a detector of MAE signals, pressure—connected to the test-pieces with vacuum grease. The transducer output is amplified by 80 dB using a pre-amplifier and a main-amplifier. The test-pieces are composed of the atomized iron powder (ATOMEL-300M) or mill-scale reduced iron powder (KIP-255M), whose average diameter distribution is shown in Fig. 2. The test-pieces were formed into a rectangular shape using several pressure at room temperature, and then sintered at about 1150°C in an atmosphere of decomposed ammonia gas. We machined them to a shape with the dimensions 9(W) × 150(L) × 9(H) mm. As for 24 kinds of test-pieces employed in this experiment, each porosity is given by the following equation.

$$P = \{(\rho_0 - \rho)/\rho\} \times 100 \tag{1}$$

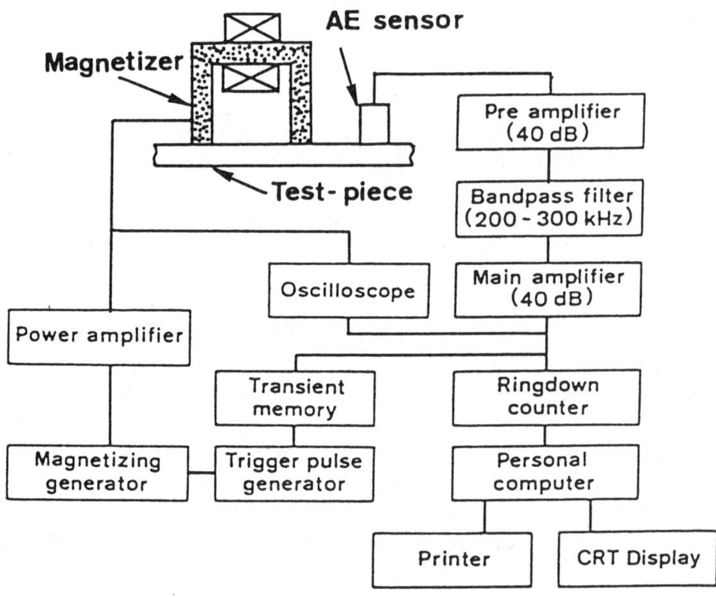

Figure 1 Schematic diagram of the experimental equipment

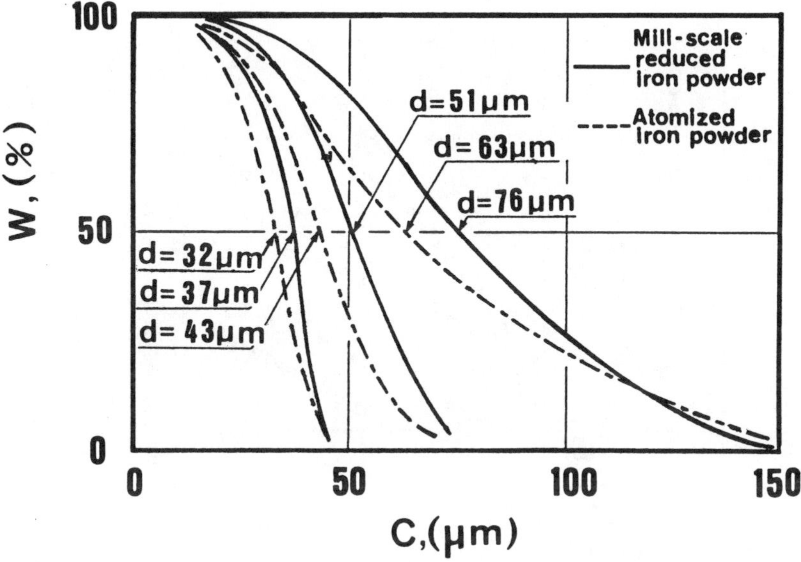

Figure 2   Relation between accumulated distribution percentage (W) and sieve opening (C)

Where ρ is the density of the test-piece and ρ_0, that of bulk material (more than 99.9% purity). Although the number and form of the pores in the sintered iron change according to the kind, average diameter and sintering condition of the iron powder, we mainly deal with the porosity having great influence on the mechanical characteristics of the sintered iron in this paper. In order to magnetize the test-pieces, we used a magnetizer with the coil around a small ferrite yoke, pressure-connected to the test-pieces. MAE is so affected by the rise time of the magnetic field that the triangle wave was used in this experiment in order to magnetize the test-pieces.

3. EXPERIMENTAL RESULTS AND DISCUSSION

The porosity shows such a favorable correlation with the mechanical characteristics of sintered iron that we have interest in the porosity as an influence factor to evaluate the material without respect to the other factors. Figure 3 shows the actual examples of measuring MAE signals produced by magnetizing the test-pieces under the magnetizing conditions of magnetizing frequency f at 60 Hz and magnetizing current I at 300 mA. And the test-pieces are made from atomized iron powder of the average diameter $d = 63$ μm. In the above figure, T shows AE count in one cycle of the magnetic-field frequency, and D shows AE count per threshold-level voltage 50 mV and the amplitude distribution of MAE. There is also an indication of MAE produced from the bulk material whose porosity is supposed to be $P = 0\%$. In accordance with the increase of the porosity of the test-pieces, both the AE count and the amplitude of MAE become greater, that is, it is recognized that the MAE changes remarkably because of the difference of the porosity. The evaluation of MAE generated in the process of magnetizing test-pieces is affected by the threshold-level voltage in the

 NOBORU SHINKE *et al.*

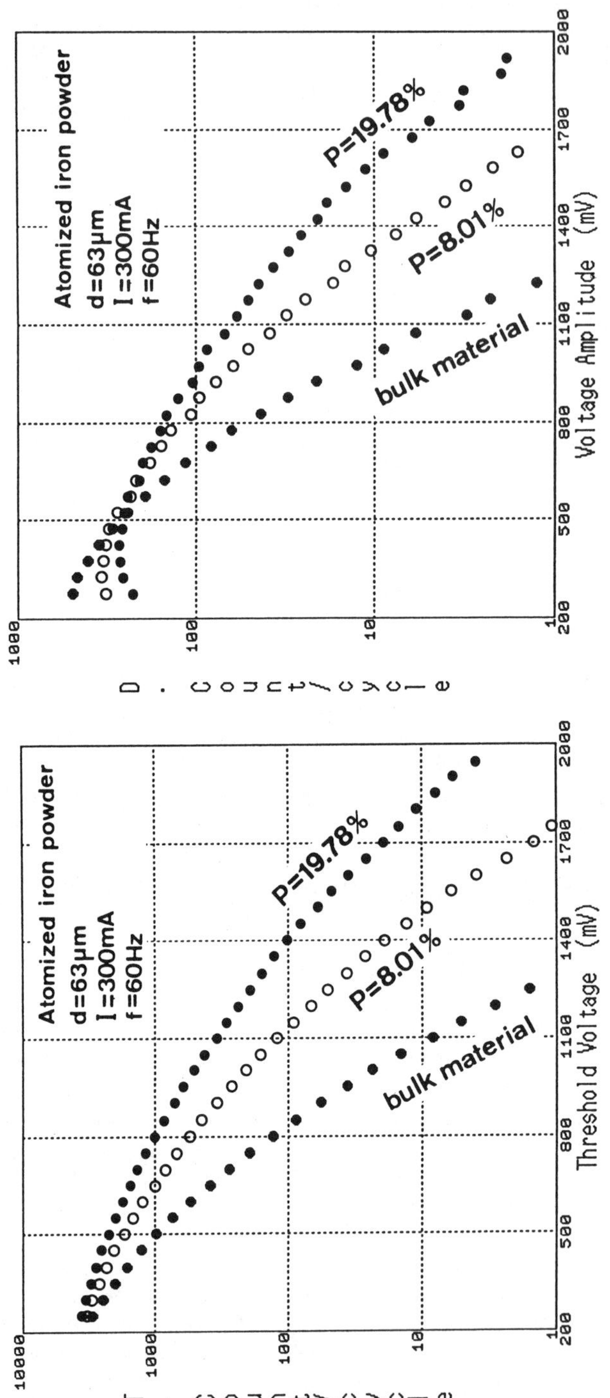

Figure 3 Measurement examples of AE count and amplitude distribution

ringdown counting method adopted in this research. The relation between the differential value D of the AE count and amplitude V_A is approximated by the following equation with constants K_1, K_2 and K_3 based on the amplitude distribution curve of MAE measured.

$$\text{Log} \cdot D = -K_1 \cdot V_A^2 + K_2 \cdot V_A + K_3 \qquad (2)$$

Where the value of K_1 shows the form of the amplitude distribution curve of MAE, which means that the amplitude of MAE produced from the test-pieces becomes smaller according to the increase of K_1. In order to investigate how the porosity of the sintered iron would affect the MAE, we measured the MAE under the magnetizing condition of $f = 60$ Hz and $I = 300$ mA, employing 24 kinds of test-pieces made from the iron powders of different average diameters. In this paper, in order to estimate the intensity of MAE signals generated from the test-pieces, we used T as AE count in one cycle of magnetizing frequency, K_1 as a constant and $\bar{e}$ as the mean event energy derived from the amplitude distribution curve. Figure 4 shows the relation between the normalized AE count $R_T = T/T^*$ (T^* for bulk material) at the threshold-level voltage of $V = 500$ mV and the porosity P of test-pieces. The equation in this figure is given by a least square method referring to the relation between the normalized AE count and the porosity, and the symbol r indicates the correlation coefficient. The normalized AE count tends to increase linearly in accordance with the increase of the porosity of the test-pieces. But the normalized AE count shows little change by the difference of the kind and average diameter of the iron powder. We

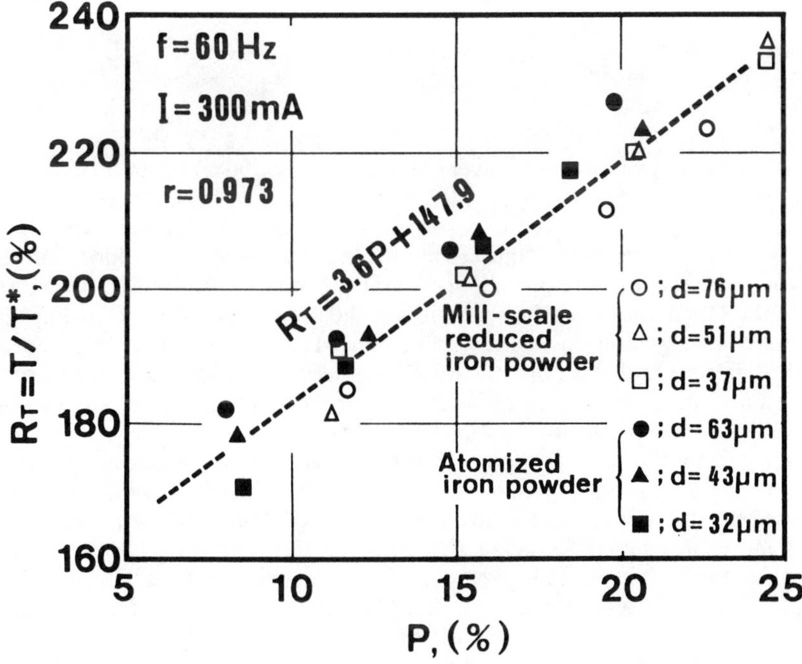

Figure 4 Relation between normalized AE count (R_T) and porosity (P)

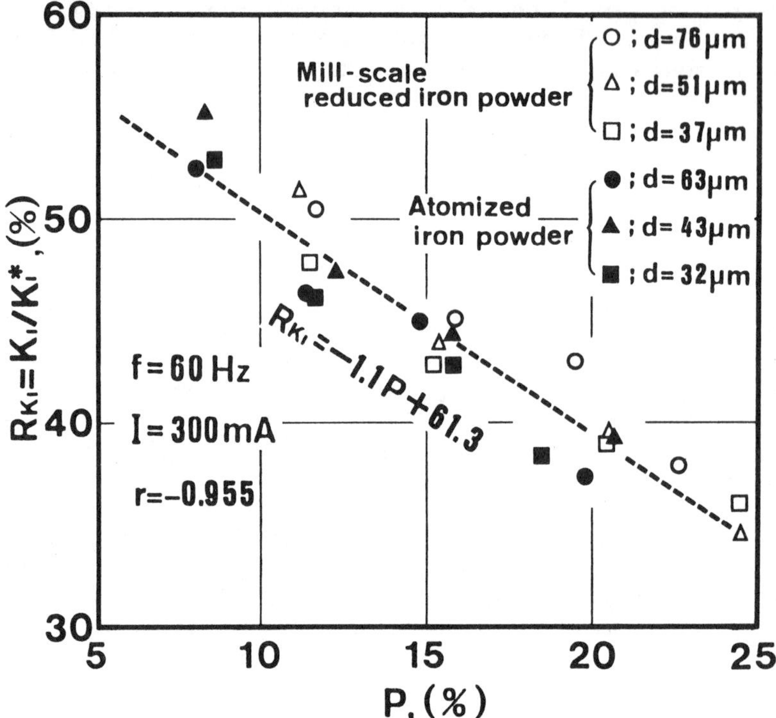

Figure 5 Relation between normalized constant (R_{K1}) and porosity (P)

measured the amplitude distribution of MAE produced by the test-pieces with different porosities in a threshold-level voltage range of 200 and 2000 mV, and found the values of K_1 by the equation (2). Figure 5 shows the relation between the normalized value $R_{K1} = K_1/K_1^*$ (K_1^* for bulk material) and the porosity P of test-pieces. As clearly seen in the figure, the normalized value R_{K1} reduces linearly in accordance with the increase of the porosity of the test-pieces. That is, we can recognize that the larger the porosity is, the larger the amplitude of MAE inclines to be. Figure 6 shows the relation between the normalized mean event energy $R\bar{e} = \bar{e}/\bar{e}^*$ ($\bar{e}^*$ for bulk material) and the porosity P of test-pieces. The results of measurements suggest an adequate correlation between the normalized mean event energy and the porosity. Taking the point of $P = 0$ on the extension of the straight line given by the least square method, the result agrees approximately with the experimental value of mean event energy for the bulk material. Now if we term β as the degree of dependence on the porosity in order to explain how it affects MAE, the mean event energy has the largest β in the parameters for evaluation of MAE. As mentioned above, it has been clarified that the porosity of the sintered iron exerts a remarkable influence on R_T, R_{K1} and $R_{\bar{e}}$ of MAE. It is generally known that the wave velocity is used as an important characteristic value to evaluate some materials, because it is sensitive to the change of the metallurgical structure. The reports have been made on the method of measuring the porosity of sintered materials nondestructively, using an adequate correlation

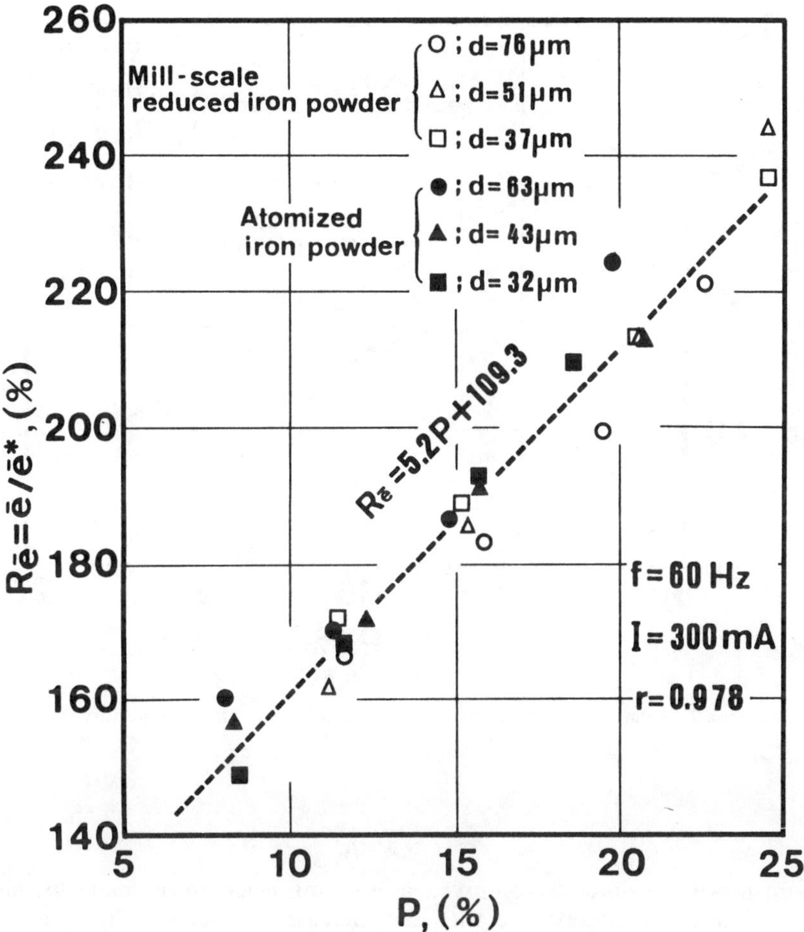

Figure 6 Relation between normalized mean event energy ($R\bar{e}$) and porosity (P)

between the porosity and the ultrasonic velocity.[7-8] Then, as for all of the test-pieces, we measured each wave velocity by the ultrasonic flaw detector made by Sonic Instruments Inc. whose transducer has a resonance frequency of about 5 MHz and an outside diameter of 1/4 inches. Figure 7 shows the relation between the normalized ultrasonic velocity $R_S = S/S^*$ (S^* for bulk material) and the porosity P of test-pieces. Within the range of porosity in this experiment, it can be considered that the effects of the porosity on the normalized ultrasonic velocity are not influenced very much by the change of the kind and the average diameter of the iron powder. Recently, M. Shibata and H. Sasaki have proposed a new model in order to explain how MAE is generated.[11] According to the model, it is understood that the intensity of MAE varies inversely with the increase of wave velocity in materials and increases in proportion to the source volume of MAE. As the skin depth of the magnetic flux H is connected with the specific resistance $\bar{\rho}$, the effective permeability $\bar{\mu}$ of the test-piece and the magnetizing frequency f, it is calculated from the following equation.

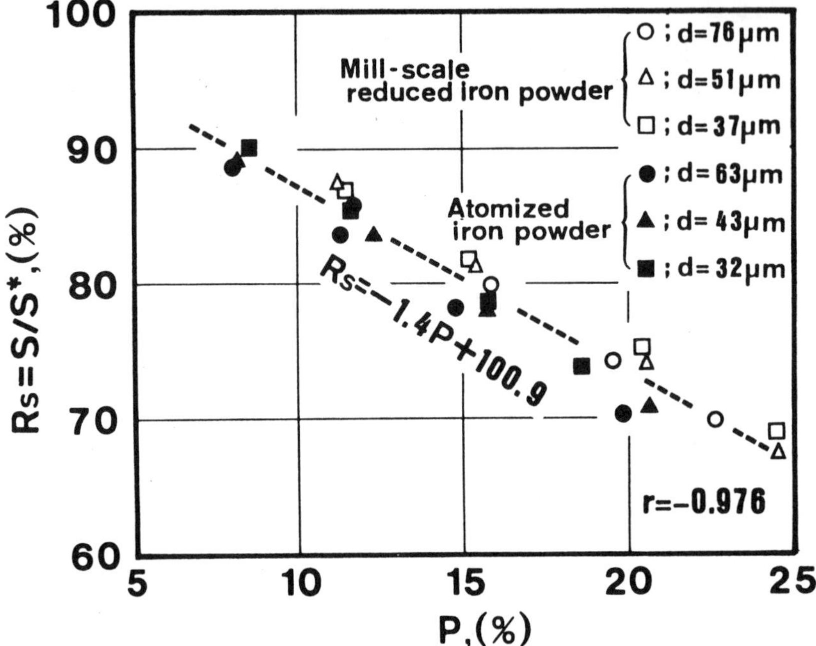

Figure 7   Relation between normalized ultrasonic velocity (*Rs*) and porosity (*P*)

$$H = \sqrt{\bar{\rho}/\pi\bar{\mu}f} \tag{3}$$

The porosity of the sintered iron exerts a great influence on the magnetic characteristics. That is, in accordance with the increase of the porosity, the specific resistance of the sintered iron becomes larger, the effective permeability smaller. Consequently, the skin depth of the magnetic flux undergoes an increase as clearly seen from Eq. (3) as well. Therefore, the source volume of MAE gets greater according to the increase of the porosity of the sintered iron. From this point of view, we can estimate the above phenomenon that the AE count and the amplitude of MAE become larger with the increase of the porosity of the sintered iron is caused by the increase of the source volume and the decrease of the wave velocity, both due to the increase of the porosity. Next, based on all of the foregoing results, we explored the feasibility of evaluating the porosity of the sintered iron by means of MAE. Table 1 indicates the values of β and r in the case of evaluating the porosity of the sintered iron by the MAE and ultrasonic velocity. The degree of dependence on the porosity is maximized by the normalized mean event energy of MAE, whose value $\beta = 5.2$ is larger than that by the ultrasonic velocity. We have hypothetically tested the difference of correlation coefficient with the significant level of 5%. Comparing the values of correlation coefficient, there is no significant difference between the MAE method and ultrasonic velocity measurement. Therefore, it has been clarified that we can nondestructively evaluate the porosity of the sintered iron by means of MAE.

Table 1. Evaluation of porosity.

Method	Parameter	β	r
M A E	R_T	3.6	0.973
	R_{K1}	−1.1	−0.955
	$R_{\bar{e}}$	5.2	0.978
Ultrasonic Velocity	R_S	−1.4	−0.976

Also, it is possible that the porosity distribution along the thickness of sintered materials is evaluated by using the skin effect.

4. CONCLUSIONS

In order to explore the possibility of nondestructive evaluation of the porosity by MAE, we have measured the MAE in the process of magnetizing the sintered iron of varied porosity. The results of this investigation are summarized as follows:

(1) The increase of the porosity of the sintered iron makes the AE count and the amplitude of MAE larger.

(2) The degree of dependence on the porosity is maximized by the mean event energy among the evaluating parameters of MAE.

(3) Irrespective of the change of the kind and average diameter of the iron powder, the porosity can be evaluated by measuring the MAE generated from the sintered iron.

The authors are also grateful to Japan Powder Metallurgy Co., Ltd. for providing the test-pieces.

REFERENCES

1) A. Sarak, V. Miskovic, E. Dudrova and E. Rudnayova: The Dependence of Mechanical Properties of Sintered Iron Compacts upon Porosity, *Powder Metallurgy International*, **6**(3), 128–132, (1974).
2) S. Kohara and K. Tatsuzawa: Effect of Porosity on the Tensile Strength of Sintered Compacts, *J. Powder and Powder Metallurgy*, **22**(4), 141–146, (1975), **23**(6), 196–200, (1976), **26**(3), 96–99, (1979), (in Japanese).
3) H. E. Exner and D. Pohl: Fracture Behavior of Sintered Iron, *Powder Metallurgy International*, **10**(4), 193–196, (1978).
4) H. Miura and Y. Tokunaga: Young's Modulus of Sintered Iron, *J. Powder and Powder Metallurgy*, **26**(2), 60–65, (1979), (in Japanese).
5) H. Miura, A. Sakamoto and Y. Tokunaga: Effect of Porosity and Pore Shape on the Young's Modulus of Sintered Iron and Steel, *J. Powder and Powder Metallurgy*, **26**(3), 82–87, (1980), (in Japanese).

6) K.S. Baird and J.D. Williams: Relationships Between Process Variables and Density of Explosively Compacted Iron Powder, *Inter. J. Powder Metallurgy & Powder Technology*, **20**(1), 23–32, (1984).
7) D. Pohl: Ermitlung der Zugfestigkeit von Sintermettallen mit Ultra-schallmessungen, *Arch. Eisenhütt.*, **40**, 647–650, (1969).
8) T. Mihara, M. Obata, T. Watanabe and H. Shimada: Nondestructive Evaluation of Porosity in Copper Powder Sintered Products by Ultrasonic Velocity Measurement, *J. JSNDI*, **37**(5), 376–380, (1988), (in Japanese).
9) A.E. Load: *Physical Acoustics*, **XI**, p. 290–353, Academic Press, New York, (1975).
10) H. Kusanagi and H. Kimura: Stress Effect on the Magnitude of Acoustic Emission during Magnetization of Ferromagnetic Materials, *J. Appl. Phys.*, **50**(4), 2985, (1979).
11) M. Shibata and H. Sasaki: MAE and BN Signals in Steel, in *Progress in Acoustic Emission*, **III**, p. 213–221, JSNDI, Tokyo (1986).
12) K. Ono and M. Shibata: Magnetomechanical Acoustic Emission of Iron and Steel, *Materials Evaluation*, **38**, 55–61, (1980).
13) N. Shinke, H. Maeda and R. Hara: Stress Effect on AE during Magnetization of SS41, *J. JSNDI*, **34**(1), 8–13, (1985), (in Japanese).
14) N. Shinke, H. Maeda and Y. Usuki: Nondestructive Measurement of the Hardened Depth by AE during Magnetization, *J. JSNDI*, **36**(4), 287–292, (1987), (in Japanese).
15) N. Shinke and H. Maeda: Temperature Effects on AE during Magnetization in a Region of Room Temperature—473K, *J. JSNDI*, **37**(9), 737-743, (1988), (in Japanese).
16) N. Shinke, H. Maeda and Y. Usuki: Development and Application of Automatic Analyzer for AE during Magnetization, *J. JSNDI*, **34**(8), 506–512, (1985), (in Japanese).

WIGNER DISTRIBUTION FUNCTION AND ITS APPLICATION TO ULTRASONIC NONDESTRUCTIVE EVALUATION*

TAKAAKI MUSHA, JUN'ICHI KAWAMURA, MICHIHIRO KITAHARA†

Acoustics Laboratory, Ono Sokki Co., Ltd. 1-16-1, Hakusan, Midori-ku, Yokohama 226 Japan

The Wigner distribution function was originally introduced in quantum mechanics as a calculation tool about half a century ago.

Recently it has attracted great interest for its validity to analyze time-varying signals, which was accomplished by the development of digital signal processing techniques.

Traditionally acoustic signals have been studied in the time or frequency domain.

The Wigner distribution function has made it possible to represent signal's power density spectrum in the time-frequency domain as a natural extension of the Fourier transformation.

In this paper, we discuss the properties of the Wigner distribution function and an example of application for the ultrasonic nondestructive evaluation is given.

KEY WORDS: Nondestructive evaluation, signal processing, ultrasonic testing, ultrasonic spectroscopy, scattering, through transmission technique, creeping wave.

1. INTRODUCTION

In the area of nondestructive evaluation (NDE), many techniques have been developed to determine the properties of materials.

Frequency analysis of ultrasonic waves propagating through a solid can provide a useful means for determining the properties of the material or characterizing the type of defects.

This method is called ultrasonic spectroscopy. It is applied for measuring the grain size in the metal, which is determined by the Fourier transformation's spectrum. However, Fourier transformation is not sufficient for analyzing the detailed structure of unsteady or transient signals such as ultrasonic scattering waves.

This is why the Fourier transformation is obtained by applying a fixed-length, moving window to the received signal prior to computing the spectrum, thus providing a trade-off between time and frequency resolution.

For such signals, the time-frequency distribution function, which is called the Wigner distribution function, can provide much clearer interpretation.

The Wigner distribution function is the most basic time-frequency distribution function and is known to have the sharpest energy resolution on the time-frequency plane without sacrificing other necessary properties. We have been working to apply the Wigner distribution function to the ultrasonic nondestructive evaluation.

---

* Originally published in Japanese in Hihakaikensa [J. JSNDI] vol. 40/No. 9 (1991)
† Faculty of Marine Science and Technology, Tokai University, Orido, Shimizu, Shizuoka 424 JAPAN.

In this chapter, basic properties are explained and an analysis example is introduced.

2. DEFINITION OF THE WIGNER DISTRIBUTION FUNCTION

The Wigner distribution function was introduced as a calculation tool to provide insights into the connection between classical and quantum mechanics by E. Wigner in 1932.

It has similar properties to the Fourier transformation in the time-frequency domain.

The Wigner distribution function of $x(t)$ and $y(t)$ is defined as[1]

$$Wxy(t,f) = \int_{-\infty}^{+\infty} x(t + \tau/2) \cdot y^*(t - \tau/2)e^{-j2\pi f\tau}d\tau \tag{1}$$

where f is for the frequency and $*$ is the complex conjugate of the signal y. If $x = y$, $Wxy(t,f)$ is called the auto-Wigner distribution function, which is expressed as

$$Wx(t,f) = \int_{-\infty}^{+\infty} x(t + \tau/2) \cdot x^*(t - \tau/2)e^{-j2\pi f\tau}d\tau \tag{2}$$

This is a function that provides a energy distribution of $x(t)$ in the time-frequency domain.

For the simplicity of the expression, we abbreviate the auto-Wigner distribution function as WDF in this paper.

With the cross product $Rx(t, \tau) = x(t + \tau/2) \cdot x(t - \tau/2)$, the WDF can be regarded as the Fourier transformation of $Rx(t, \tau)$ in the τ-axis. The τ-axis is an independent axis from t-axis, therefore the WDF is not affected directly from the uncertainty relation of time and frequency resolutions. This provides the WDF a high resolution of ultrasonic waves. Practical calculation of $Wx(t,f)$ can be obtained by using the FFT algorithm for the sampled data of $x(t)$.

For its computation, the sampling rate must be faster than the frequency multiplied four by the upper frequency of the signal, or the band-pass filter, which cutting frequency is below the quarter of sampling frequency, must be used after sampling at the rate of the frequency twice as the upper frequency of the signal.

Examples that show the high resolution of the WDF spectrum are shown in the following figures.

Figure 1 represents the FFT's spectrogram of the chirp signal, which is a linear frequency-modulated signal changing with time. Figure 2 represents the WDF spectrum of the same signal. It is evident that the WDF has a higher time-frequency resolution than the spectrogram which is obtained by FFT using the fixed-length moving window.

3. USEFUL PROPERTIES OF THE WDF

The WDF has a number of useful properties including the ability to provide high time-frequency resolution for certain classes of nonstationary signals.[2]

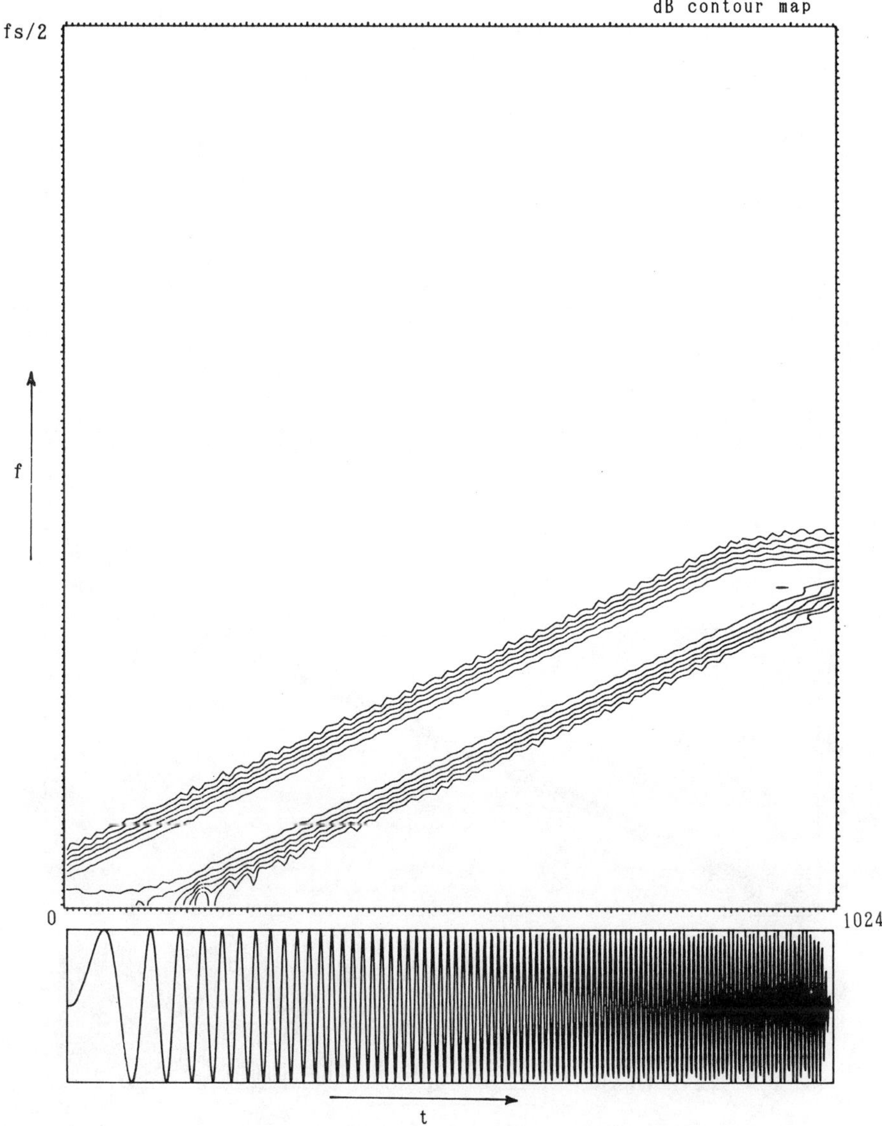

Figure 1   Spectrogram of chirp signal (dB contour map)

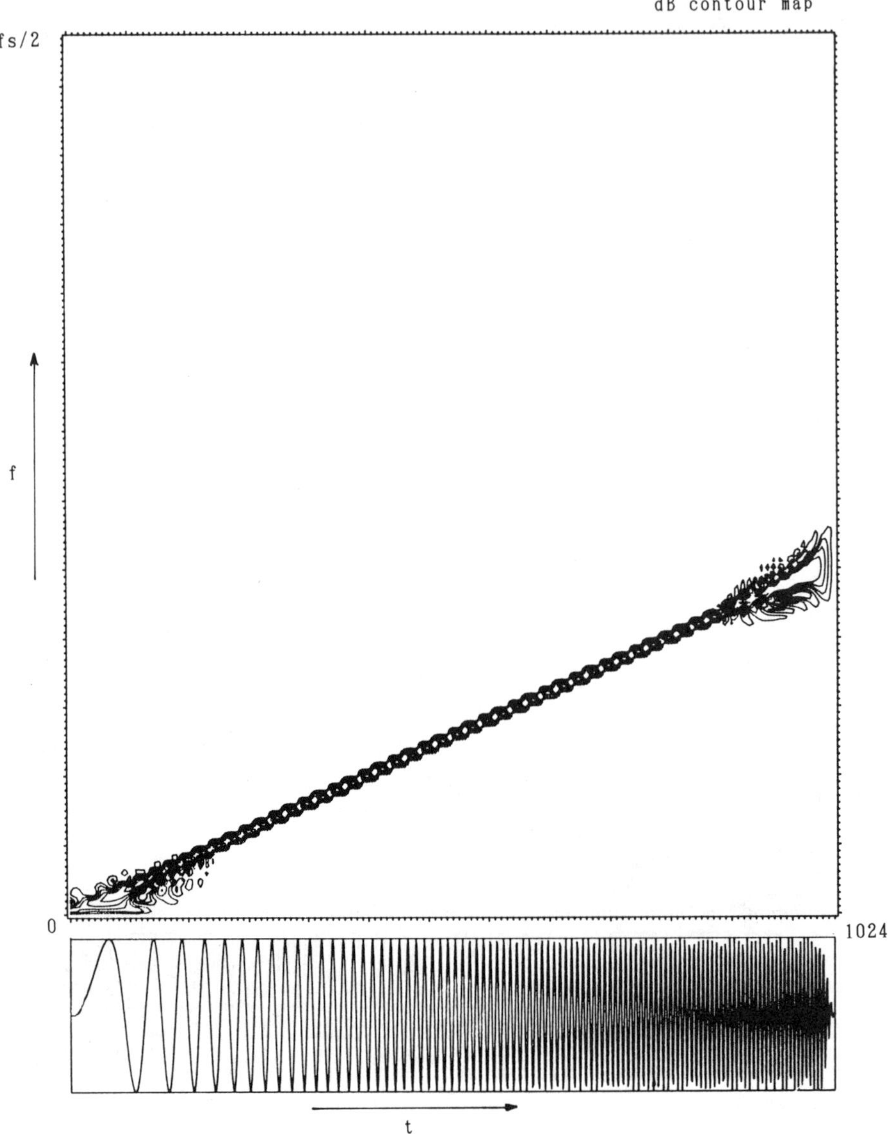

Figure 2 WDF of chirp signal (dB contour map)

(P1) The integration of the WDF with respect to the frequency yields the instantaneous power.

$$\int_{-\infty}^{+\infty} Wx(t,f)df = |x(t)|^2$$

(P2) The time integration of the WDF yields the power density spectrum

$$\int_{-\infty}^{+\infty} Wx(t,f)df = |X(f)|^2$$

(P3) A time shift of the signal has the corresponding shift in the WDF

$$\text{If } y(t) = x(t-t_0) \text{ then } Wy(t,f) = Wx(t-t_0,f)$$

(P4) A frequency modulation of the signal has the corresponding shift in the WDF.

$$\text{If } y(t) = x(t)e^{j2\pi f_0 t} \text{ then } Wy(t,f) = Wx(t,f-f_0)$$

(P5) The WDF has real value.

$$\text{Im}[Wx(t,f)] = 0$$

(P6) If the signal $x(t) = 0$ for $t < T1$ and $t > T2$, then $(t,f) = 0$ for $t < T1$ and $t > T2$.

(P7) If $X(f) = 0$ for $f < F1$ and $f > F2$, then $Wx(t,f) = 0$ for $f < F1$ and $f > F2$.

(P8) The first-order moment of the WDF with respect to time equals to group delay.

$$\int_{-\infty}^{+\infty} t \cdot Wx(t,f)df \bigg/ \int_{-\infty}^{+\infty} Wx(t,f)dt = t_g(f)$$

(P9) The first-order moment of the WDF with respect to frequency gives the instantaneous frequency.

$$\int_{-\infty}^{+\infty} f \cdot Wx(t,f)df \bigg/ \int_{-\infty}^{+\infty} Wx(t,f)df = f_i(t)$$

Each of these properties has a specific significance in signal analysis. Especially the property P9 is effective for the ultrasonic spectroscopy, which is related to the evaluation of the waveform parameter to recover the material properties.

Conventionally material property is determined by the signal spectrum, which is given by

$$f_m = \int_{f_1}^{f_2} f \cdot G(f)df \bigg/ \int_{f_1}^{f_2} G(f)df \tag{3}$$

where $G(f)$ is a power spectrum of the signal. The property P9 makes it possible to estimate the centroid of the spectrum at any time of interest, so that we can observe signals with different time delays and frequency composites.

Contrary to these useful properties, the WDF has undesirable characteristics.[3] An undesirable characteristic is the occurence of cross terms in the WDF of multi-component signals arising from the interaction of the individual time-different and frequency-different components. Namely the WDF of the sum of two signals $x(t)$ and $y(t)$ is expressed as

$$Wf + g(t,f) = Wf(t,f) + Wg(t,f) + 2\text{Re}[Wfg(t,f)] \qquad (4)$$

where $Wfg(t,f)$ is the cross-WDF that contributes components at the mean of the frequency components of f and g.

Another undesirable characteristic of the WDF is the occurence of the negative values. The occurence of negative values is consistent with the Heisenberg's uncertainty which prohibits the arbitrary sharp frequency discrimination. But these undesirable characteristics can be cancelled by smoothing the WDF spectrum over a region of the time-frequency plane, where the time-smoothing and frequency-smoothing are usually chosen independently.

4. APPLICATION FOR THE ULTRASONIC SCATTERING WAVE ANALYSIS

Scattering from a defect provides the interesting problem to determine the shape and physical property of the defect.[4] The WDF appears to be a very useful tool to characterize these scattering waves.

Figure 3 shows specimens with three types of defects, cylindrical hole, spherical void and crack, prepared for measurements. The specimens are made of concrete. Figure 4 shows the measuring method of forward-scattered elastic waves. The incident pulse was emitted from the transducer on one side of the specimen and received by the transducer on the other side of specimen.

To cancel the reflections and scatterings except from the defects, the process shows in the Figure 5 was performed for the receiving signals. $N(f)$ is a Fourier transformation of the reference signal $n(t)$, which is the wave through the specimen with no defects. $X(f)$ is the Fourier transformation of the forward-scattering wave $x(t)$.

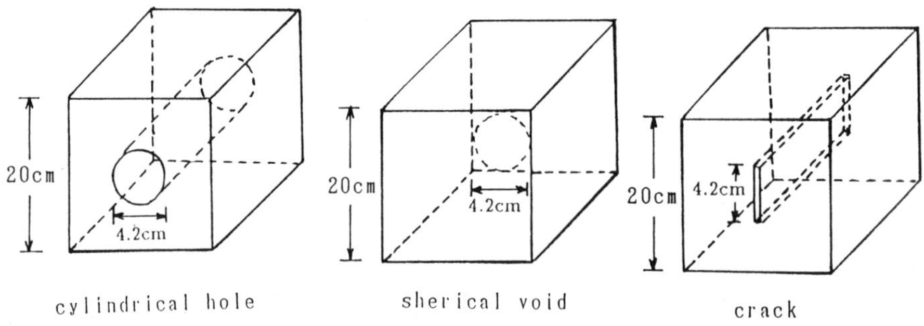

Figure 3　Types of defect

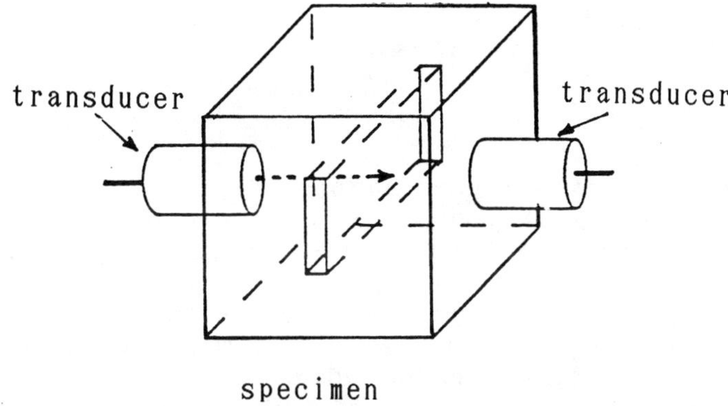

Figure 4 Measuring method of forward-scattering value

n(t): Signal with no defect

x(t): Signal with defect

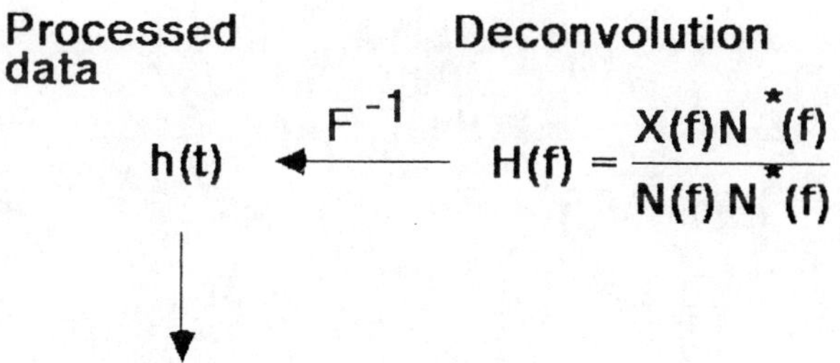

$$W_h(t, f) = \int_{-\infty}^{\infty} h(t+\tau/2)h^*(t-\tau/2)e^{-j2\pi f\tau}\,d\tau$$

Figure 5 Processing diagram for computation of $Wh(t, f)$

Processed signal $h(t)$ was obtained through the deconvolution and the inverse Fourier transformation shown as follows.

$$h(t) = F^{-1} \left[\frac{X(f) N^*(f)}{N(f) N^*(f)} \right] \tag{5}$$

where F^{-1} denotes the inverse Fourier transformation.

Figures 6–8 show the WDF spectrum for the processed signal $h(t)$ for cylindrical hole, crack and spherical void, respectively. For cylindrical hole, the WDF spectrum is periodic in time direction.

This means that scattering waves generated by the creeping waves around the cylinder is predominant for cylindrical type of defect (Figure 6). For spherical void, almost of all energy quickly transmit forward (Figure 7).

For crack, the combined effect of the edge diffraction and creeping waves can be recognized (Figure 8). It is also recognized that the dominant frequency for crack is rather high compared with those for cylindrical hole and void. It is interesting to quantify the data in details by using the WDF properties of P8 and P9.

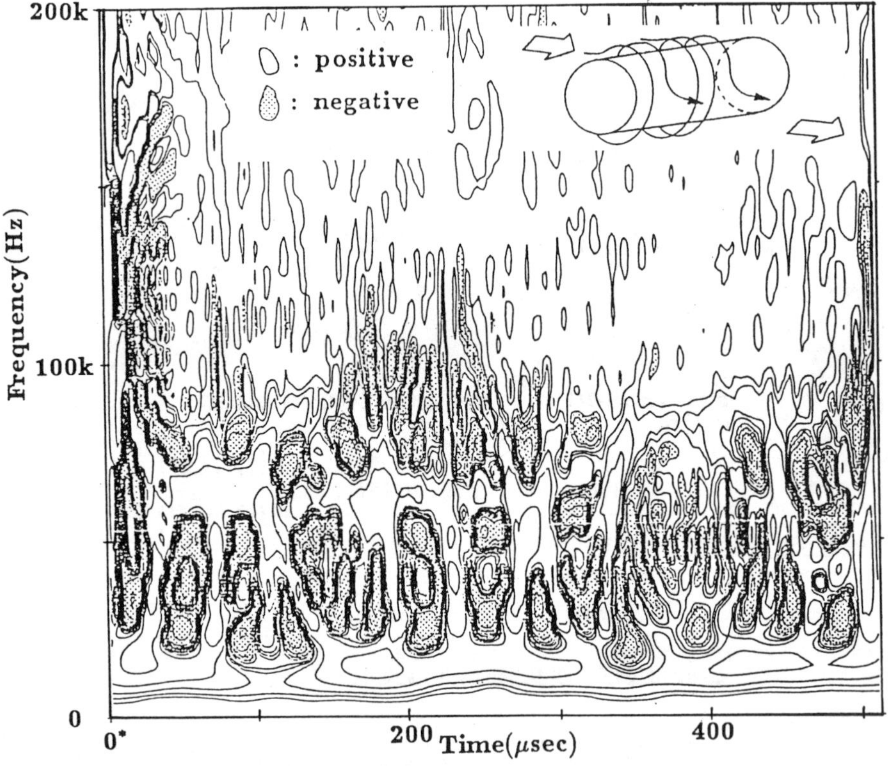

Figure 6   WDF of processed signal for cylindrical hole

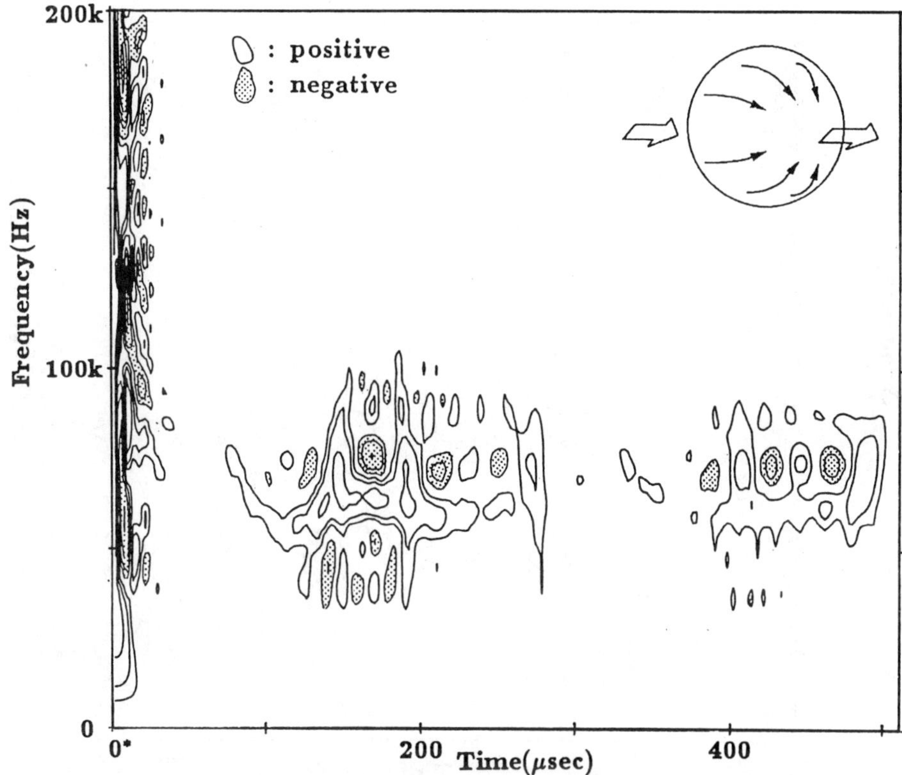

Figure 7   WDF of processed signal for spherical hole

5. CONCLUSIONS

In the examples discussed here, the WDF gives us a effective method to analyze many features associated with the scattering waves in solid. The WDF spectrum gained from such analysis of the scattering waves also provides a useful understanding of scattering mechanism. However, the WDF has undesirable characteristics of the occurence of cross terms and negative values on the time-frequency plane. Moreover, the WDF requires many components for computation and it takes such time to complete the WDF algorithms by the conventional computer. But these disadvantages might be cancelled by using the adequate signal processings and high-speed processors.

The application for the WDF of the scattering waves in solid appears to be a promising means for extracting information from the ultrasonic scatterings for composites or other materials. This method will become a powerful tool for the ultrasonic nondestructive evaluation in the future.

REFERENCES

1) T. A. C. M. Classen and W. G. F. Necklenbrauker: The Wigner Distribution—a tool for time-frequency signal analysis, Parts I, II and III, *Philips J. Res.*, **35**, (1980).

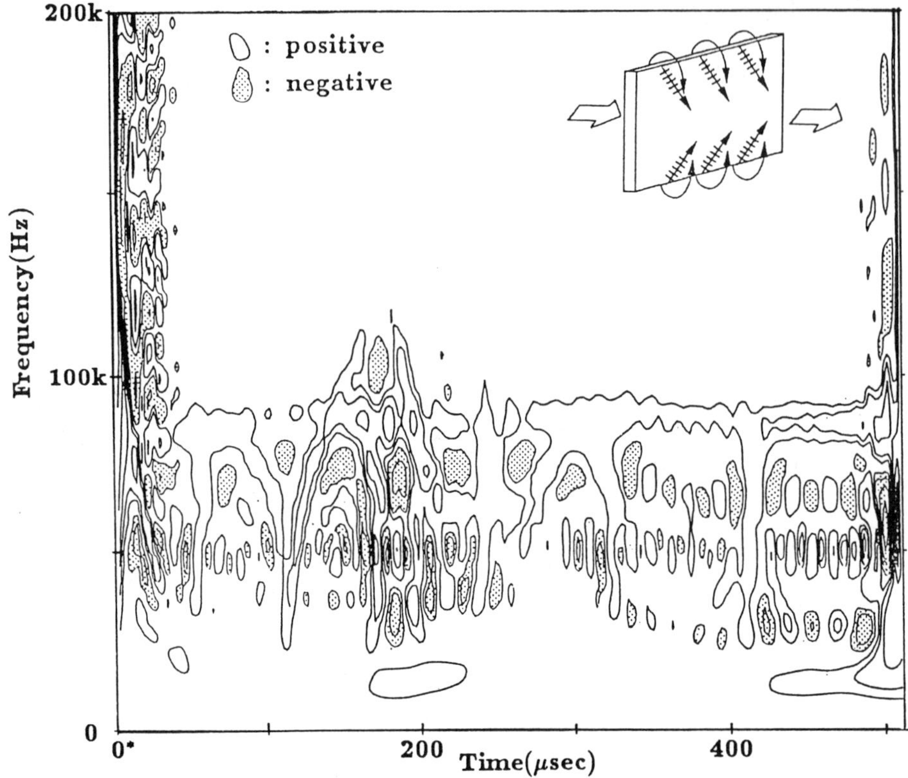

Figure 8   WDF of processed signal for crack

2) C.P. Janse and A.J.M. Kaizer: Time-Frequency Distribution of Loud-speakers: The Application of the Wigner Distribution. *J. Audio Eng. Soc.*, **31**, 195–223, (1983).

3) F. Hlawatsch and W. Krattenthaler: A New Approach to Time-Frequency Signal Decomposition, *Proc. IEEE, ISCAS '89*, 1248–1251, (1989).

4) M. Kitahara, T. Musha and J. Kawaura: Wigner Distribution Applied for Scattering Characterization of Defect in Concerete, IEEE 1990 *Ultrasonics Symposium Proceedings*, (1991).

5) N. Yen: Time and Frequency Representation of Acoustic Signals by means of the Wigner Distribution Function: Implication and Interpretation, *J. Acoust. Soc. Am.*, **81**(6), 1841–1850, (1987).

6) N. Yen, L.R. Dragonette and S.K. Numrich: Time-Frequency Analysis of Acoustic Scattering from Elastic Objects, *J. Acoust. Soc. Am.*, **87**(6), 2359–2370, (1990).

ESTIMATION OF IMPACT DAMAGE IN CFRP LAMINATES WITH COMPLIANCE METHOD*

TADAHARU ADACHI, KEI SAKANOUE†, SADAYUKI UJIHASHI, HIROYUKI MATSUMOTO

Department of Mechanical Engineering, Tokyo Institute of Technology, 2-12-1 Oh-okayama, Meguro-ku, Tokyo, 152, Japan

The estimation of impact damage in CFRP (Carbon Fiber Reinforced Plastics) laminates with the compliance responses of the deflection is suggested. By the instrumented drop weight impact test, the strain histories in the drop weight are measured and converted to the impact loads and deflections according to one-dimensional wave equation. It is obtained that the impact load histories with damage include the signals having high frequencies. The CFRP laminates have the same compliance responses whenever impact damage does not occur. When the impact damage generates or develops, the amplitude of the compliance response increases. The maximum value of the compliance response becomes larger with the development of the impact damage.

KEY WORDS: Composite Material, Impact Damage, Drop Weight Test, Measurement, Compliance.

1. INTRODUCTION

The reduction of compression strength after impact (CAI) for carbon fiber reinforced plastics (CFRP) on behalf of advanced composite material is an important problem which must be solved. The estimating technique of impact damage must be established for the damage tolerant design. It is not adequate to apply the damage estimation method for isotropic metallic material to that of fiber reinforced plastic (FRP) laminates. The failure pattern of FRP is different from that of metallic material. For example internal damages, such as delaminations and transverse cracks of FRP laminates are larger than damages on external faces. Then, the instrumented measuring system are necessary to estimate damages of FRP laminates.

Lateral impact test by drop weight is mainly used to estimate impact strength of FRP. The drop weight impact test has the advantages of the application to actual structure testing and no enforcement of fracture pattern, comparing with Charpy test or Izod test which are classified as the pendulum type. The object of practical drop weight impact tests in NASA Standard[1] and ASTM standard[2] for plastics is the resistance of penetration phenomena. And in these standard tests, impact load histories, load-deflection diagrams and absorption energy consumed during generating damages are adopted as criteria in design of FRP structures[3-5]. For CAI problems, the behavior of damages by sub-perforation velocity is important. As the energy for generating such damages is little, the new accurate technique for measure-

---

* First published in Japanese in Hihakaikensa [J. JSNDI] vol. 40/No. 3 1991
† Kei Sakanoye was a post graduate student at the Tokyo Institute of Technology. He now works for the Central Japan Railway, the address of whom is 1-6-5 Marunouchi, Chiyoda-ku, Tokyo 100, Japan.

ment is needed to obtain the variation of load and deformation in time by generating damage and to be established to estimate damage of FRP.

In the present paper, we propose that the compliance responses of CFRP laminates are adopted to estimate impact damage by using the instrumented measuring system of drop weight test. As impact loads varies according to dimensions of a drop weight and a specimen, materials of them and impact velocity, generally, each impact load measured by experiment cannot be compared directly to obtain the characteristics of the specimen. The impulsive characteristics of FRP laminates under the same condition of the load are evaluated by using the compliance responses. The compliance response is defined as the deflection at the impact point of the specimen subjected to a unit load. It is evaluated based on the load and deflection histories in the experiment. Componeschi and Stinchcomb[6] select the stiffness variations of FRP specimen as the index of damage estimation in fatigue experiment. The initiation, propagation and extent of impact damage will be estimated with the compliance responses in this paper.

Loads and deflections in time are computed accurately, considering stress wave propagating in a bar as a drop weight. The authors[7] have already certified that accurate measurement is yielded by the present method. And it was obtained that the signal having high frequencies on the impact load are related to damage generation of FRP laminate[7]. In the present paper, it is found out that time of crack initiation coincides with that of appearance of sharp signal in the load history. The variation of the compliance responses for each impact to CFRP laminates is investigated.

2. FUNDAMENTAL THEORY

2.1. *Principle of Measurement for Load and Deflection*

Let us consider that a circular bar with a strain gauge falls on the FRP laminate specimen with the velocity V_0 as shown in Figure 1. Dynamic loads and deflections at the end of the bar are computed from the strain histories of the bar. In Figure 1, l and l_1 are the length of the bar and the position where the strain gauge is affixed.

The waves propagating in the bar are governed by the following one-dimensional elastodynamic theory.

$$\frac{\partial^2 u}{\partial x^2} = \frac{1}{C^2} \frac{\partial^2 u}{\partial t^2},$$ (1)

$$C^2 = E/\rho,$$

where u, x and t are the displacement, the axial coordinate of the bar and time, respectively. And E and ρ are Young's modulus and density of the bar.

Using Laplace transform with the origin of the time $t = 0$ as the instant when the bar collides on the specimen, the general solution of Eq. (1) is given by the equation:

$$\bar{u}(x, p) = A_1 \cdot \exp(px/c)$$
$$+ A_2 \cdot \exp(-px/C) - V_0/p^2$$ (2)

where A_1 and A_2 are unknown coefficients determined with the boundary conditions, and

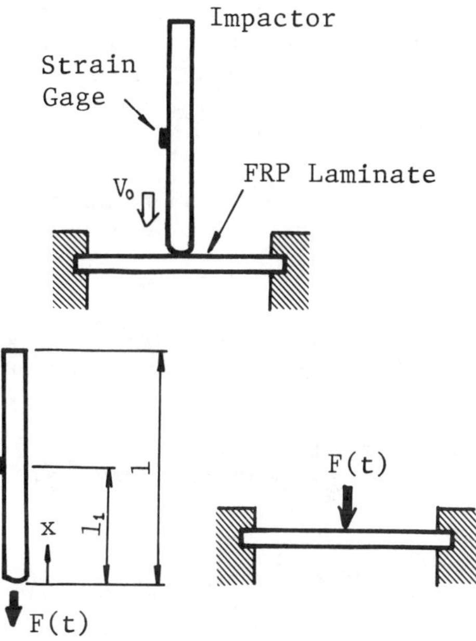

Figure 1   Principle of measurement

$$\bar{u}(p) = \int_0^\infty u(t) \cdot \exp(-pt)dt$$

When the upper end of the bar is free and $\epsilon_1(t)$ is the strain history measured by the gauge at $x = l_1$, the boundary conditions arc

$$\left. \begin{array}{l} \text{at } x = l \ \sigma_x(= E\partial u/\partial x) = 0 \\ \text{at } x = l_1 \ \partial u/\partial x = \epsilon_1(t) \end{array} \right\} \tag{3}$$

The coefficients A_1 and A_2 of Eq. (2) are determined from Eq. (3).

When the strain gauge is affixed on the center point of the bar i.e. $l_1 = l/2$, the Laplace transformed load $\bar{F}(p)$ can be inverted analytically. The load in time applied on the specimen are derived as the following closed form:

$$F(t) = AE\{\epsilon_1(t+l/2C) \cdot H(t+l/2C) + \epsilon_1(t-l/2C) \cdot H(t-l/2C)\} \tag{4}$$

where A and $H(t)$ are the cross-sectional area and unit step function, respectively. Therefore the load can be computed by experimental data $\epsilon_1(t)$ shifted in time.

During the period when the bar is contacted with the specimen, the deflection of the specimen is coincided with the displacement at the lower end of the bar. Then, the Laplace transformed deflection $\bar{w}(p)$ is obtained as

$$\bar{w}(p) = \frac{C}{p} \frac{\bar{\epsilon}_1(p) \cdot \cosh(pl/C)}{\sinh(pl/C)} + \frac{V_0}{p^2} \tag{5}$$

In this case, the analytical formula of the deflection is not given in a simple closed form. Then the deflection in time is obtained directly from transformed deflection

$\overline{w}(p)$ expressed with $\overline{\epsilon}_1(p)$ by the numerical method explained in the next section.

The compliance response of the specimen is defined as the following form using the load and deflection in transformed domain.

$$\overline{R}(p) = \frac{\overline{w}(p)}{\overline{F}(p)} \frac{1}{p} \tag{6}$$

The compliance response is yielded by the numerical inverse Laplace transform, too.

2.2. *Numerical Laplace Transform*

The numerical procedure using fast Fourier transform[8,9] is applied to the Laplace transform and inversion for Eqs. (5) and (6). The formulation for pair of numerical Laplace transform are expressed as follows:

$$\overline{f}_n = \frac{T}{N} \sum_{k=0}^{N-1} f_k \exp(\gamma k \Delta t - i 2 \pi n k / N)$$

$$\tag{7}$$

$$f_k = \frac{\exp(\gamma k \Delta t)}{T} \sum_{n=0}^{N-1} \overline{f}_n \exp(i 2 \pi n k / N)$$

where

$$f_k = f(k\Delta t), \overline{f}_n = \overline{f}(\gamma + i n \Delta \omega), \Delta \omega = 2\pi/T,$$

$$\Delta t = T/N, k = 0, 1, 2, \ldots, N-1, i = \sqrt{-1}$$

and Δt and N which indicate sampling period and sampling number respectively, must be coincided with experiment conditions. In this study, following values are taken in Eq. (7): $\gamma = 4/T$, $\Delta t = 1 \, \mu s$ and $N = 4096$.

3. MEASURING SYSTEM

Figure 2 shows the measuring system of impact load-deflection in time for FRP laminated specimen. A steel bar with 5 mm in radius and 600 mm in length falls freely to impact at the center of CFRP laminates. The CFRP specimen having square dimension (180 mm × 180 mm) is held between two steel plates having a circular hole with 150 mm in diameter and thickness of 8 mm. At the longitudinal center of the bar, a pair of semi-conductor strain gauges (Kyowa, KSP-2-E4) are affixed symmetrically about its center axis and connected in series to cancel the bending moment components. The analogue data of strain gauges which are produced by the bridge circuit, are memorized in transient digital analyser (Autonics, APC-204, memory: 12 bits and 4096 words). The sampled data which are transferred to a microcomputer (NEC, PC-9801VX), are converted to the load-deflection in time according to Eqs. (4) and (5). The impact velocity is obtained with the universal counter (Iwatsu, SC-7201) by measuring the duration when a barrier affixed on the bar cuts a laser beam just before the collision on the specimen. The barrier has 5.4 mm in thickness. The low-pass filter was never used for all experimental and computational results in the present paper.

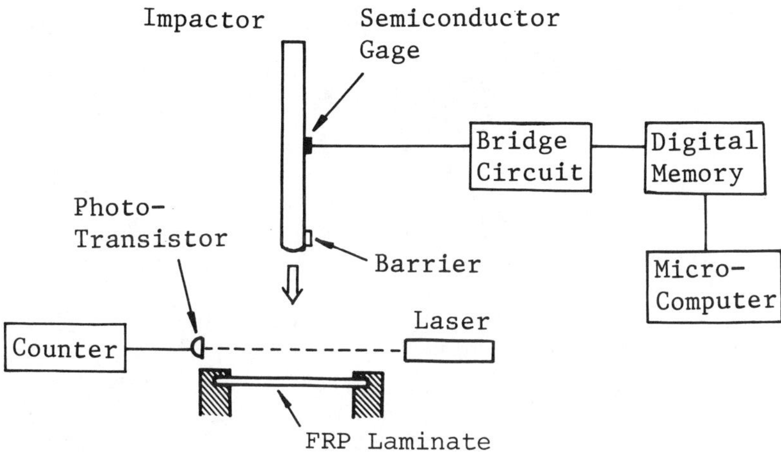

Figure 2 System of instrumented falling drop weight impact testing

4. DETECTING OF DAMAGE SIGNAL ON IMPACT LOAD HISTORY

The authors[7] have obtained that impact load histories include signal having high frequencies for generation of impact damages in CFRP laminates. From the experimented data, we will consider when the signal appears. When the CFRP laminate is impacted, a fiber-directional crack is generated on the backside and also the delaminations are initiated on interfacial boundaries generally. Takeda et al.[10] has shown the crack and the delaminations in GFRP laminate are generated within the period of approximate by 30 μs. We compare the appearance times of damage signals on the load history with the generation time of the crack on the backside of the CFRP plate. The specimen was laminated as $[0°_2/90°_2/-45°_2/45°_2]_{SYM}$ by using uni-directional prepreg (Toray, P3051–15). The crack gauge (Kyowa, KV-5) was bonded on the backside of the specimen just under the impact point. By measuring the resistance variation of the crack gauge, the appearance time of the crack could be decided. When the steel bar collided on the virgin CFRP specimen with the velocity of 3.2 m/s, the load in time was measured. After this experiment, the impact delaminations on six interlaminars was actually certified with scanning acoustic microscope (Olympus, UH3, Frequency of Acoustic Lens: 30 MHz). Figure 3 shows the load history and the variation of output voltage from the bridge circuit as the results of resistance variation of the crack gauge. From Figure 3, the appearance time (A) of damage signal on the impact load is coincided with the time of the discontinuous point (B) on the output history from the crack gauge. As a result, it is obtained that the damage signal appears when the damage is generated.

5. ESTIMATING IMPACT DAMAGE WITH COMPLIANCE RESPONSES

The variation of compliance responses is investigated, when iterative impacts are applied to a CFRP laminated plate. The stacking sequence of the prepreg in the

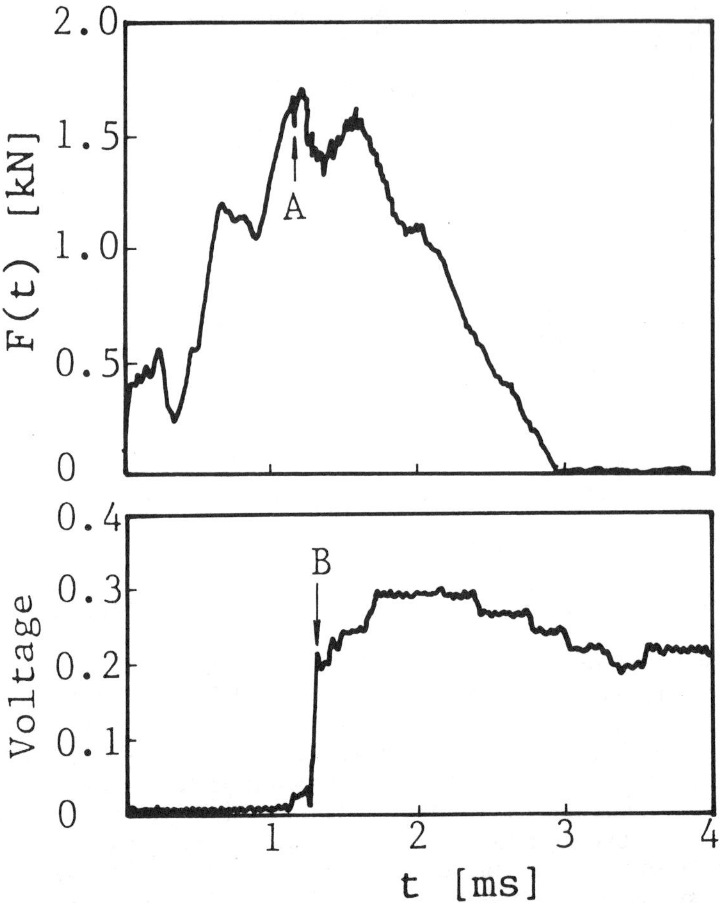

Figure 3   Relation between load history and output from crack gauge

specimen was $[0°_4/90°_4]_{SYM}$ with the same prepreg as one used in Chap. 4. The feature of the observed crack on the backside and the appearance of damage signal on the load for each applied impact on the same specimen are summarized in Table 1. When the No. 4, 8 and 9 impacts were applied to the specimen, it could be found that the damages generated or developed by the observation of the cracks on the external faces and the damage signals on the load histories. The load histories of No. 3, 4 and 5 are shown in Figure 4. The solid line and the broken line indicate the cases of generation of damage during impact process, and no generation or no development, respectively. Only by comparing with each load history, it is difficult to recognize the variations of the impulsive characteristics of laminates before and after generation of damage.

The compliance responses are computed based on load histories and deflection histories in all experiments. Figure 5 shows the typical compliance responses at the cases of no damage (No. 1 and No. 3) and generating damage during impact process

Table 1. The results of the experiment.

NO.	IMPACT VELOCITY [m/s]	EXISTANCE OF DAMAGE SIGNAL ON IMPACT LOAD	CRACK ON BACKSIDE OF SPECIMEN
1	1. 99	NO	—
2	2. 23	NO	—
3	2. 70	NO	—
4	3. 45	YES	GENERATION
5	2. 09	NO	NO DEVELOPMENT
6	2. 73	NO	NO DEVELOPMENT
7	3. 31	NO	NO DEVELOPMENT
8	3. 81	YES	LITTLE DEVELOPMENT
9	5. 21	YES	LARGE DEVELOPMENT
10	2. 33	NO	NO DEVELOPMENT

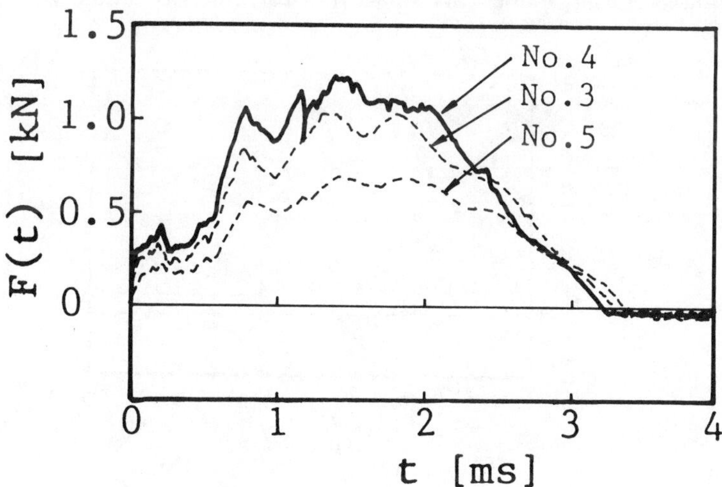

Figure 4 Impact load histories (————: generation of damage during impact process, -----: no generation or development of damage)

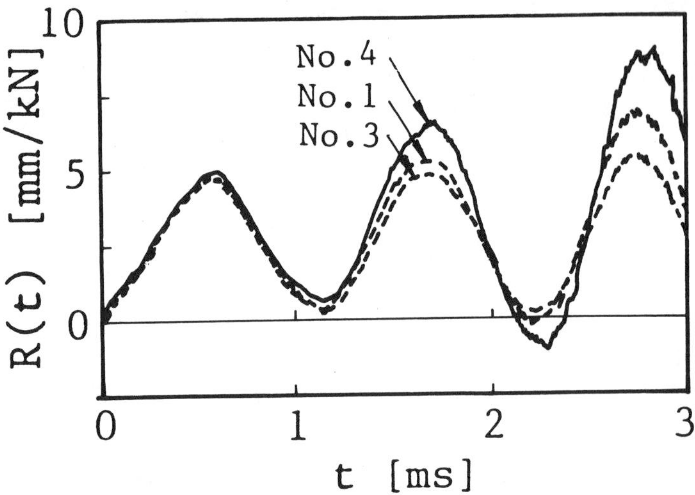

Figure 5 Compliance responses of deflection at the impact point (————: generation of impact damage, – – – –: before generation of impact damage)

(No. 4). The compliance response of No. 1 approximately coincided with the ones of No. 3 and No. 2 which is not shown in this Figure. Though the compliance response of No. 4 is the same as the others just after the collision between the bar and the specimen, the amplitude of response of only No. 4 much increases with the lapse of time. For the sake of more detailed consideration of the results in Figure 5, the variations of the extreme values $(R_2 - R_1)/R_1$ in all compliance responses are plotted in Figure 6, where R_1 and R_2 are the extreme values of first and second period, respectively. The reason of estimating the responses until the second period is that the computation error using numerical Laplace transform increases rapidly after the third

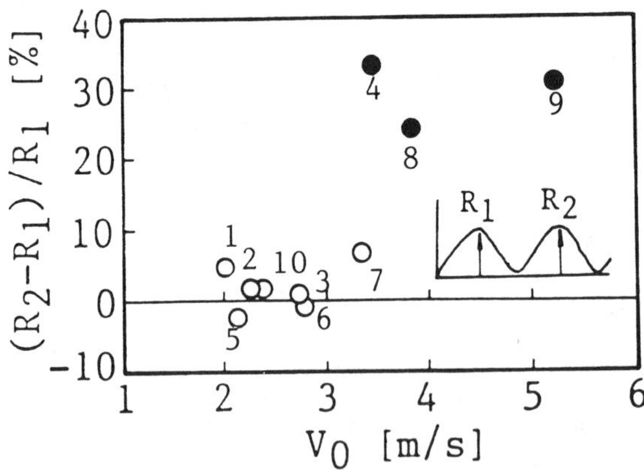

Figure 6 Increasing rates of extreme values of compliance responses (○: generations or developments of impact damage, ●: no generation or development of impact damage, each number indicates order of impact.)

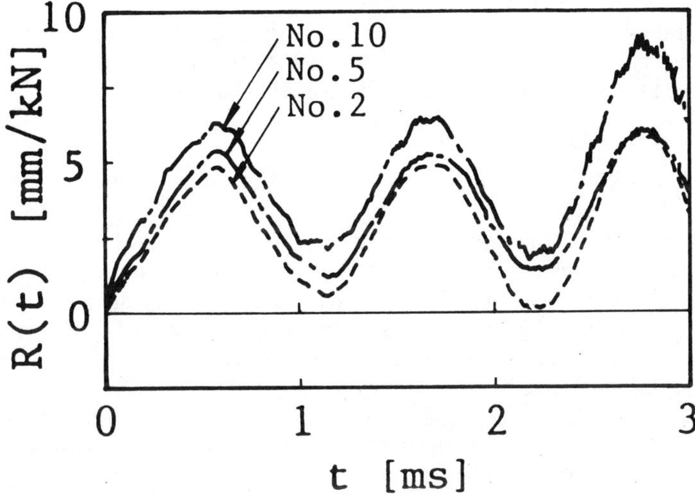

Figure 7 Compliance responses of deflection at the impact point (-----: before generation of impact damage, ———: after generation or development of impact damage)

period at this experiment case. In Figure 6, ● and ○ indicate the existence and absence of damage signal on the load history, respectively. It is shown that $(R_2 - R_1)/R_1$ in generating or developing the damage are much larger than ones with no damage or no developing. Therefore the variation of the compliance responses is caused by the stiffness reduction of the specimen during the damage increasing.

Let us compare among the compliance responses of No. 2 before generating the damage, No. 5 after generating and No. 10 after developing in Figure 7. As shown in Figure 7, the amplitude of responses increase by generation and development of the damage. Then, the relation between the impact velocity V_0 and the extreme value R_1 of the first period in the compliance response is illustrated with Figure 8. The marks of ○

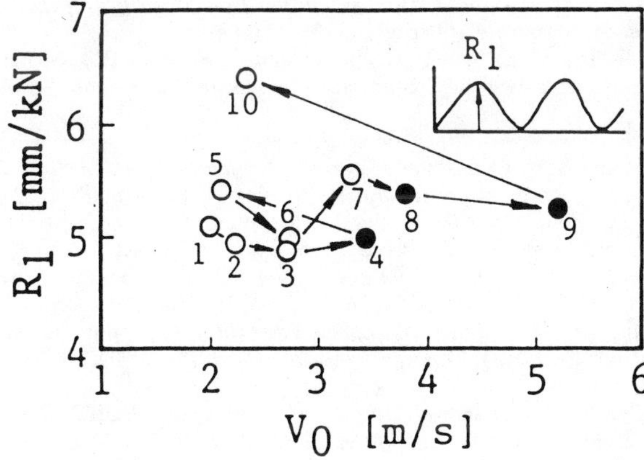

Figure 8 First extreme values of compliance responses (○: no generation or development of impact damage, ●: generation or development of impact damage, each number indicates order of impact)

and ● in Figure 8 have the same meanings as in Figure 6. Figure 8 shows the tendency that the R_1 values of ○ become larger whenever the damage generates or develops in the specimen. This variation of R_1 values is due to the stiffness reduction of the specimen by generating and developing the impact damage.

6. CONCLUSIONS

The estimating method for impact damage of CFRP laminates was investigated by using the drop weight testing system presented in this paper. The impact loads and the deflections of the specimens in time could be measured accurately with consideration of the stress wave propagation in a bar as the drop weight. From two obtained histories, the compliance responses at the impact point were computed. It was obtained that the impulsive characteristics of the CFRP laminates could be evaluated easily with the compliance responses which were independent on the magnitudes of the impact loads. As the results in the present experiment, it was shown that the CFRP laminates have the same compliance responses whenever the impact damage does not occur. When the impact damage generates or develops, the amplitude of the compliance response increases. The maximum value of the compliance response becomes larger with the development of the impact damage.

The present investigation are performed during the tenure of Grant-in-Aid for General Scientific Research (63420027) of the Ministry of Education, Science and Culture. The authors would like to thank for this grant.

REFERENCES

1) *Standard Tests for Toughened Resin Composites, RP-1092*, NASA, 1–35, (1983).
2) Standards Test Methods for Impact Resistance of Rigid Plastics Sheeting or Parts by Means of a Tup (Falling Weight), *ASTM Standards and Literature References for Composite Materials*, ASTM, New York, 456–471, (1987).
3) J. D. Winkel and D. F. Adams: Instrumented Drop Weight Impact Testing of Cross-Ply and Fabric Composites, *Composites*, **16**(4), 268–278, (1985).
4) P. O. Sjoblom, J. T. Hartness and T. M. Cordell: On Low-Velocity Impact Testing of Composite Materials, *Journal of Composite Materials*, **22**, 30–52, (1988).
5) B. Z. Jang, L. C. Chen, C. Z. Wang, H. T. Lin and R. H. Zee: Impact Resistance and Energy Absorption Mechanisms in Hybrid Composites, *Composite Science and Technology*, **34**, 305–335, (1989).
6) E. T. Camponeschi and W. W. Stinchcomb: Stiffness Reduction as an Indicator of Damage in Graphite/Epoxy Laminates, *Composite Materials: Testing and Design Sixth Conference*, (Ed. by I. M. Daniel), *ASTM STP787*, American Society for Testing and Materials, 225–246, (1982).
7) S. Ujihashi, K. Sakanoue, T. Adachi and H. Matsumoto: Experimental Measurement of the Mechanical Properties of Fibre-Reinforced Plastics under Impact Loading, *Proceedings of the Mechanical Engineerings Forth International Conference FRC'90*, IMechE, 157–162, (1990).
8) T. Adachi, K. Hatano, S. Ujihashi and H. Matsumoto: Impulsive Responses of Framed Structures Using a Matrix Method with a Numerical Laplace Transform (1st Report, A Case of Structures Consisting of Straight-Bar Elements), *Transactions of the JSME (Series A)*, **56**(724), 917–923, (1990), (in Japanese).
9) T. Adachi, S. Ujihashi, H. Matsumoto: Impusive Response of a Finite Circular Cylindarical Shell Subjected to Waterhammer Waves, *Bulletin of JSME*, **29**(249), 737–742, (1986).
10) N. Takeda, R. L. Sierakowski, C. A. Ross and L. E. Malvern: Delamination-Crack Propagation in Ballistically Impacted Glass/Epoxy Composite Laminates, *Experimental Mechanics*, **22**, 19–25, (1982).

ENERGY FLOW IN REMOTE FIELD EDDY CURRENT TECHNIQUE

HIROSHI HOSHIKAWA and KIYOSHI KOYAMA

College of Industrial Technology, Nihon University, 1-2-1 Izumicho Narashino Chiba 275, Japan

Remote field eddy current technique (RFECT) has recently been considered as a promising in-service inspection method for ferromagnetic tubing. However, its principle is so different from the conventional eddy current testing that some phenomena have been remaining obscure.

The authors have made the quantitative investigation of the electromagnetic energy flow in RFECT to clarify the phenomena underlying the technique. The energy generated by the exciter decays very rapidly inside the tube and the energy getting outside the tube travels with a small decay rate along the tube axis. As a result, the energy flow intensity is larger outside the tube than inside and the remote field area is dominated by the energy penetrating through the tube wall from outside of the tube. Thus RFECT can detect defect in ferromagnetic tubing because the energy containing the information on the tube wall exists in the remote field area.

KEY WORDS: NDI, In-service inspection, Ferromagnetic tube, Eddy current testing, Remote field eddy current technique, Electromagnetic energy flow.

1. INTRODUCTION

A larger number of metal tubing have been used in power plant, chemical plant and so forth. Those tubings are liable to get degraded during their long service although they were installed perfect at the construction. Thus periodic in-service inspections are required for them in order to ensure the safety of the structures. In-service inspection of tubing has to be conducted from tubing inside because of its structural configuration. On top of that, it has to be done at a certain speed because the inspection has lots of tubing to be finished during very short period of those facilities' suspension. The conventional eddy current test method using inner probe has successfully been used for the inspection of non-magnetic tubing. However, the method has not been successful to inspect ferromagnetic tubing because magnetic permeability of tubing has a significant influence on the skin effect and its irregularity also causes large noises in the inspection. Magnetic saturation technique does not work effectively in this case. Thus ferromagnetic tubing has been uninspectable and required the advent of new inspection techniques.

Remote field eddy current technique (RFECT) was originally developed as a in-service inspection method for oil well casing in 1950s.[1-2] However, it did not attract any attention of the people in NDT circle because its principle is so different from the conventional eddy current testing that people suspected the technique. The publication by T. R. Schmidt made a debut of the technique in 1984 coupled with the needs of a new inspection technique for ferromagnetic tubing. Since then, RFECT has been studied as an inspection method of oil pipe lines in Canada.[4-5] Three major gas companies in Japan have developed the inspection technique for gas distribution tubing by using the technique.[6-9] RFECT has also been used as the in-service inspection method for the heat exchanger tubing in nuclear power plant.[10] The first international conference on remote field technique was held in 1988. While RFECT has started to attract attention of people in NDT circle, the details of phenomena underlying the technique have not been clarified thoroughly yet. The authors already reported the magnetic flux line distribution to elucidate the phenomena,[11-15] but not all of them have been clarified.

In this paper, the authors report the quantitative electromagnetic energy flow in RFECT in order to clarify the phenomena.[16-17] It is derived that some of the energy generated by the exciter manages to penetrate through the tube wall, travel axially along the tube outer surface, and penetrate the wall again back inside the tube at the position displaced more than two tube diameters away from the exciter. Thus RFECT can detect defects in ferromagnetic tube from its inside because electromagnetic field in remote field area contains information on the tube wall.

2. REMOTE FIELD EDDY CURRENT TECHNIQUE

The conventional eddy current method using inner coil detects the impedance variation of the exciter coil or the voltage variation of detector coil near the exciter. Thus conventional method utilizes the change of magnetic flux near the exciter. The area near the exciter is called near field (NF) area.

On the other hand, RFECT picks up the variation of the electromotive force induced in the detector axially displaced about two tube diameters away from the exciter as shown in Figure 1. Thus RFECT utilizes the variation of magnetic flux inside the tube in the remote field (RF) area which is displaced more than two tube diameter from the exciter.

T.R. Schmidt has explained the RFECT phenomenon as follows.[3] The electromagnetic field in NF area is dominated by the direct coupled field from the exciter and decays very rapidly as the distance increases from the exciter. On the other hand, the electromagnetic energy outside the tube which manages to penetrate through the tube wall travels a long distance along the tube axis without decaying much. The energy outside the tube becomes larger than inside at RF area. Some of the energy outside the tube penetrates again through tube wall back inside the tube. As a result, the electromagnetic field at RF area is dominated by the energy penetrating through the tube wall. Thus the electromagnetic field in RF area contains the information on the tube wall and can be used to detect defects in the wall. T.R. Schmidt has composed the qualitative explanation based on experimental results. However, the explanation is not enough when we need to determine test parameters such as the test frequency, the arrangement of test coils, and so forth. The authors have derived the electromagnetic energy flow in RFECT in order to elucidate remote field phenomena quantitatively.

3. ELECTROMAGNETIC ENERGY FLOW

The flow of electromagnetic energy is derived from Poynting vector. The time average flow of harmonic electromagnetic field is determined by the real part of complex Poynting vector P given by the following equation.

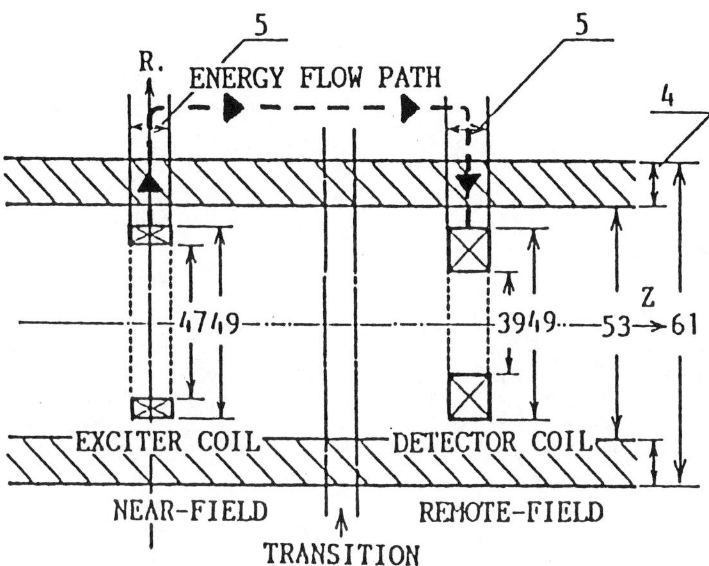

Figure 1 Remote field representation [unit:mm]

$$P = E \times H^* \tag{1}$$

where E is electric field and H^* is conjugate of magnetic field and $\times$ indicates vector product.

The authors have first derived vector potential A in RFECT by finite element method in the case of axisymmetrical geometry, then derived electric and magnetic fields by using following relations.

$$E = -j\omega A \tag{2}$$

$$H = 1/\mu\{-\mathbb{R}\,\partial A/\partial z + \mathbb{Z}(\partial A/\partial r + A/r)\} \tag{3}$$

where the symbols are as follows. (r, z): cylindrical coordinate, ω: angular frequency, μ: magnetic permeability, $j = (-1)^{1/2}$, $\mathbb{R}{:}r$ unit vector, $\mathbb{Z}{:}z$ unit vector. The time average energy flow is derived by substituting equations (2) and (3) into equation (1).

Figure 1 shows the analyzed geometries of the ferromagnetic tube, exciter, and detector. Since the flow of energy has both its intensity and direction, the authors have adopted triangle marks $\triangle$ in order to show both of them. Since the intensity varies incomparably depending on the position, the lengths of the marks have been determined proportional to the value $k_1 \log_{10}(k_2 P)$ where P is given by equation (1) and k_1 and k_2 are appropriate constants. The current density in the exciter is assumed $0.3 \times 10^7\,\mathrm{A/m^2}$ throughout the finite element analysis in this paper.

4. ENERGY FLOW DISTRIBUTION

Figure 2 shows electromagnetic energy flow distributions at various test frequencies. It is seen that the energy generated by the exciter is absorbed into inner tube surface

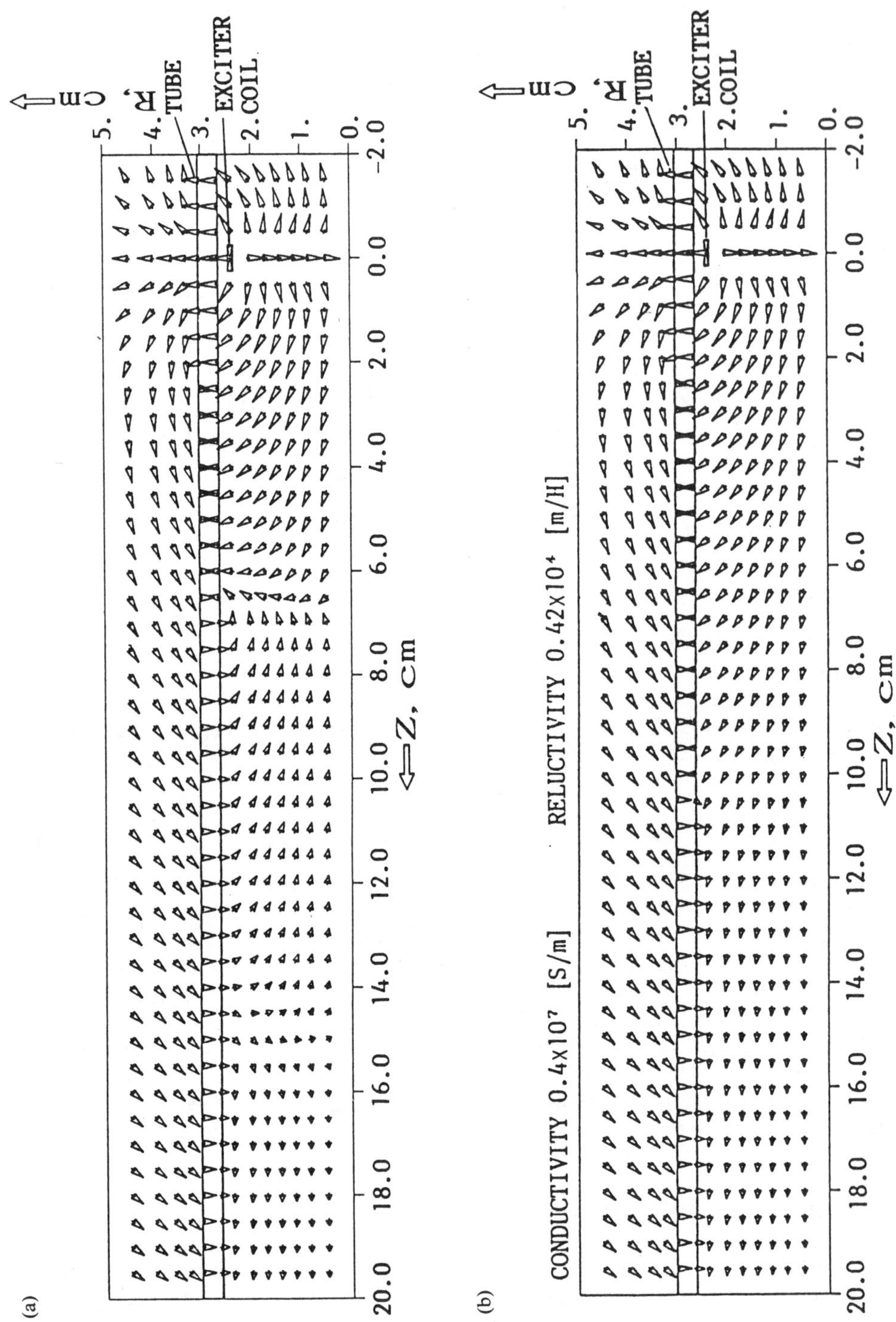
5. R, cm
4. TUBE
3.
EXCITER
2. COIL
1.
0.
-2.0
0.0
2.0
4.0
6.0
8.0
10.0
12.0
14.0
16.0
18.0
20.0
=Z, cm
(a)
CONDUCTIVITY 0.4x10⁷ [S/m]
RELUCTIVITY 0.42x10⁴ [m/H]
(b)

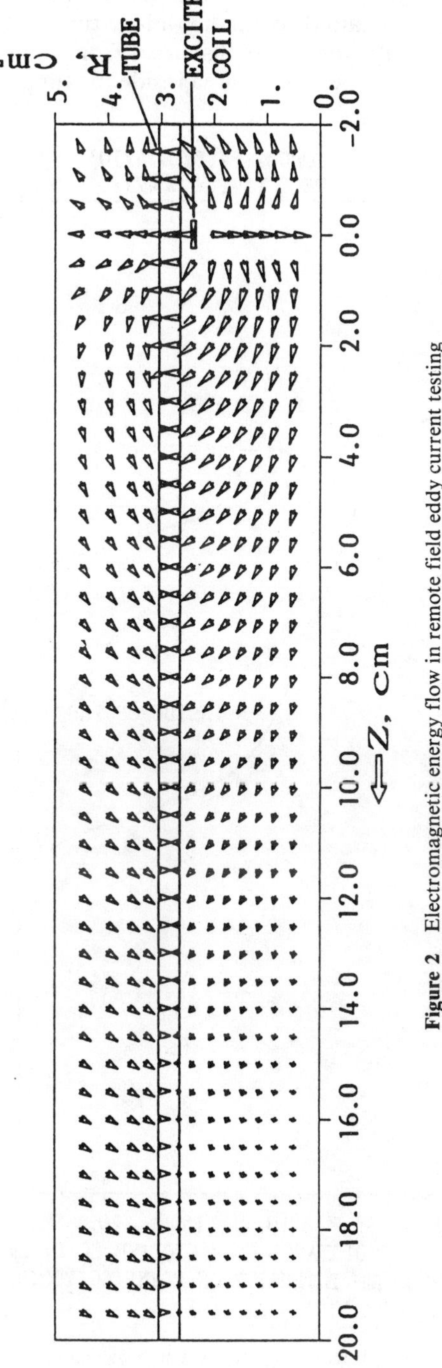

Figure 2 Electromagnetic energy flow in remote field eddy current testing
(a) Frequency: 80 Hz
(b) Frequency: 120 Hz
(c) Frequency: 400 Hz

and decays in the tube wall because it is dissipated as eddy current loss. It is also shown that some of the energy manages to penetrate through the wall, travel along tube axis outside the tube, and penetrate again through the wall getting back inside the tube at the axial position about two tube diameters displaced from the exciter. The fact is verified by the figure (b) because there is the area inside the tube where energy flows

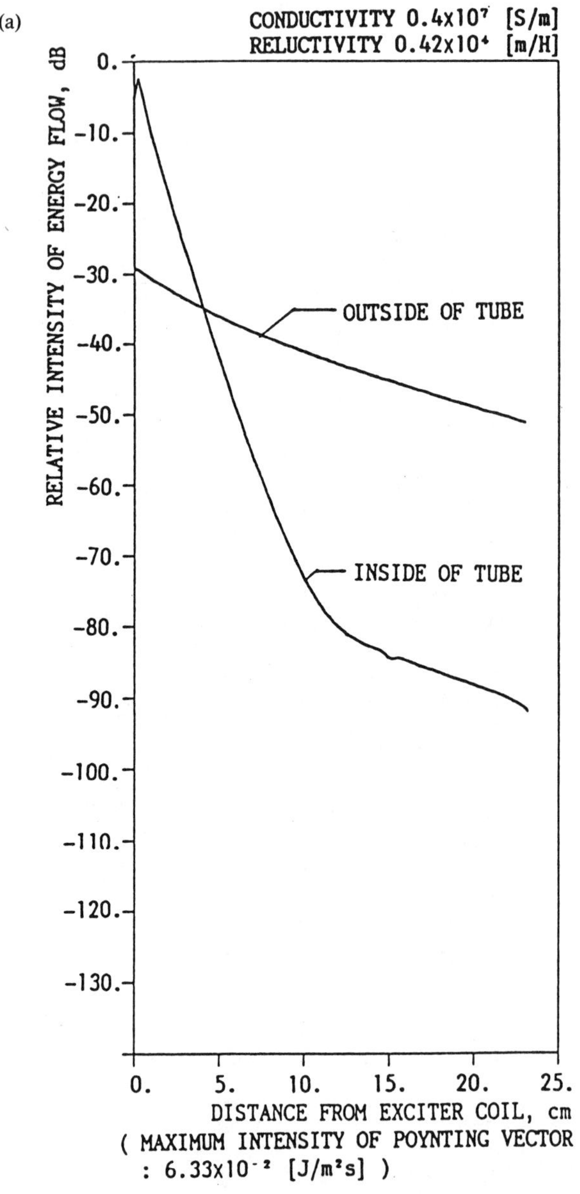

Figure 3 Energy flow intensity in inside and outside of tube
(a) Frequency: 80 Hz

even toward the exciter.

Figure 3 shows the variations of energy flow intensity at the radial coordinates 2.5 mm apart from both inner and outer tube surfaces as functions of axial distance from the exciter. The intensity inside the tube decays very rapidly in the NF area, the area near the exciter, while it does not decay much in the RF area, the area more than two tube diameters displaced from the exciter. On the other hand, the flow intensity

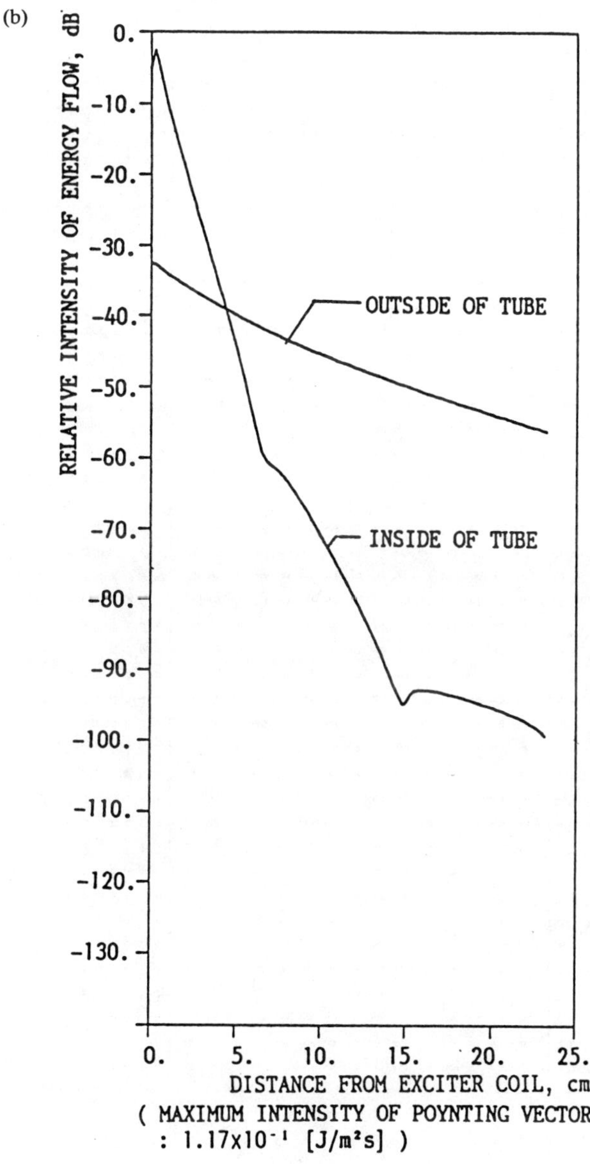

Figure 3 (b) Frequency: 120 Hz

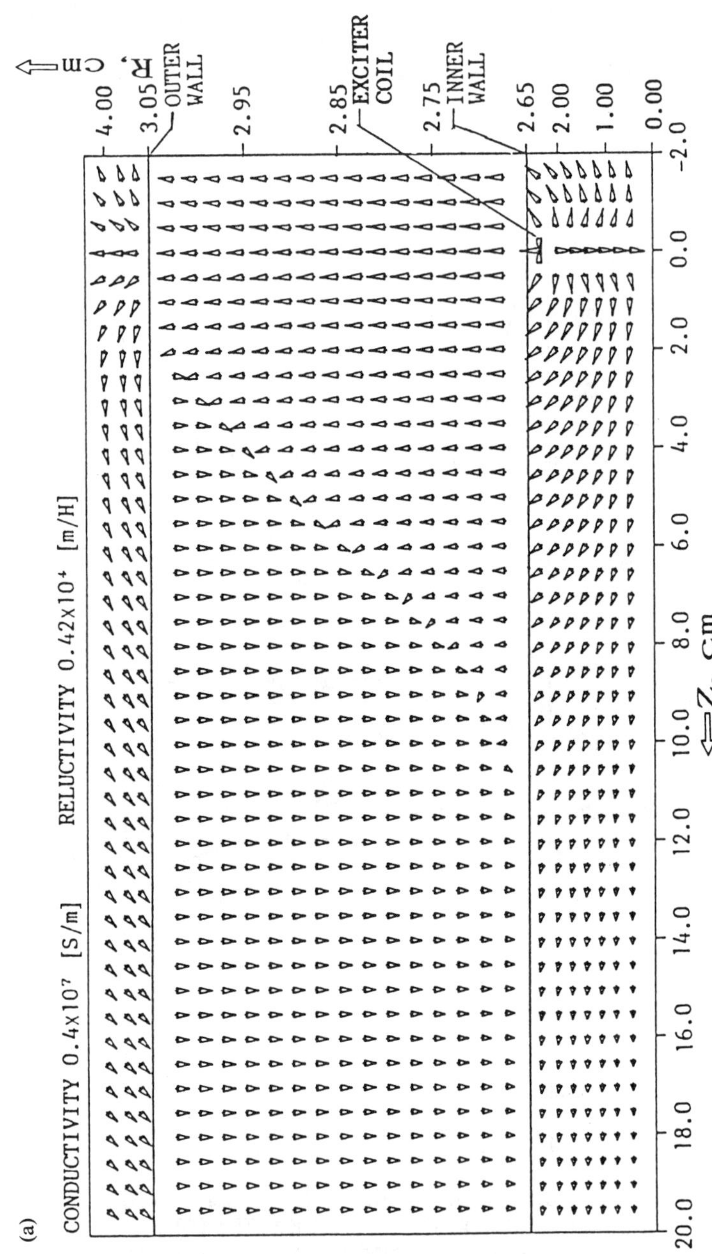

CONDUCTIVITY 0.4x10⁷ [S/m]
RELUCTIVITY 0.42x10⁴ [m/H]
R, CM
4.00
3.05
OUTER WALL
2.95
2.85
EXCITER COIL
2.75
INNER WALL
2.65
2.00
1.00
0.00
Z, cm
-2.0
0.0
2.0
4.0
6.0
8.0
10.0
12.0
14.0
16.0
18.0
20.0
(a)

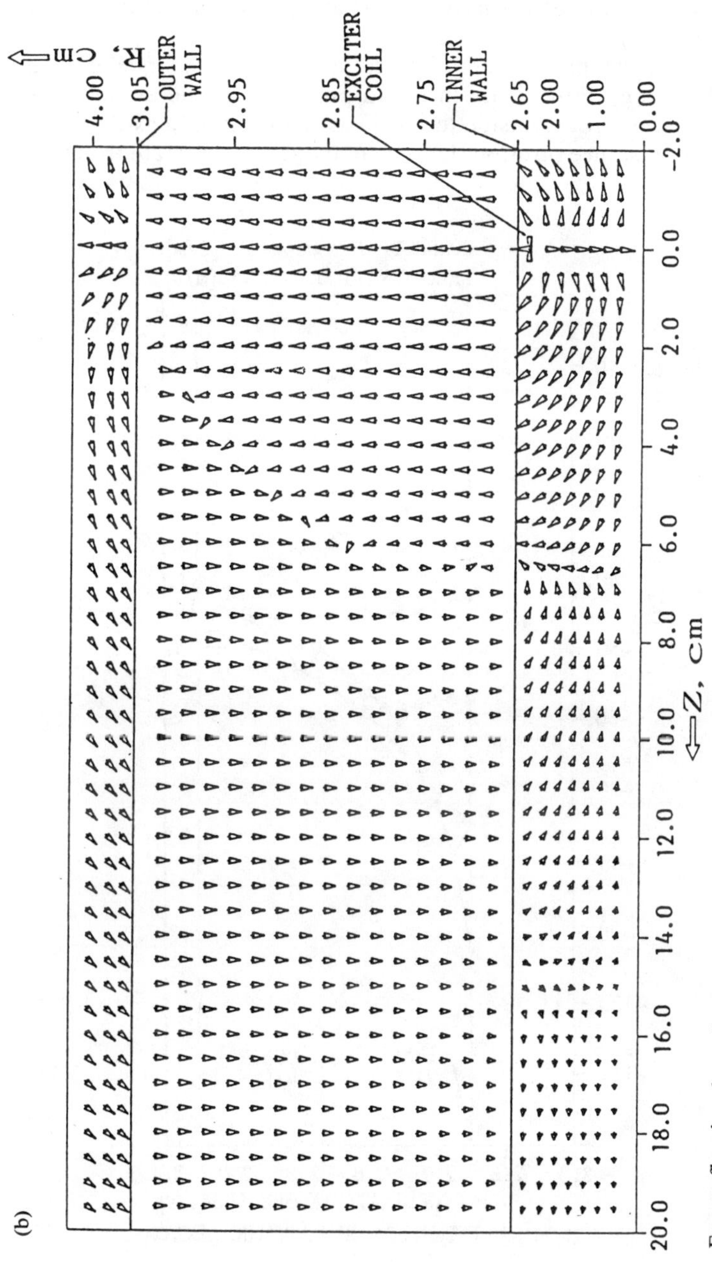

Figure 4 Energy flow in tube wall
(a) Frequency: 80 Hz
(b) Frequency: 120 Hz

outside the tube does not decay much. Thus the intensity is larger outside the tube than inside in the RF area. It is seen that the curve indicating the intensity inside the tube has two distinct changes of curvature. The positions of these curvature changes are found to coincide with the positions where the energy flow direction changes inside the tube in Figure 2(b).

Figure 4 shows the details of energy flow in the tube wall enlarged only the radial coordinate. The energy flows from inside to outside near the exciter. The energy begins trying to travel from outside to inside as the position gets distant from the exciter. At farther distance position the flow manages to penetrate through the wall

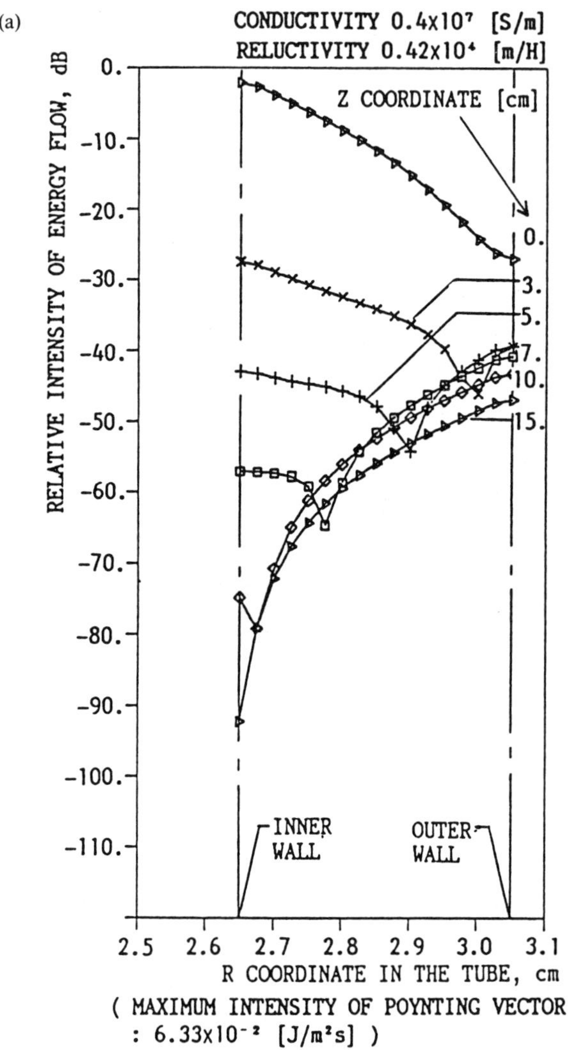

Figure 5 Energy flow intensities in tube wall as functions of radial coordinate (a) Frequency: 80 Hz

back into tube inside. Electromagnetic energy is consumed in the tube wall as eddy current loss.

Figure 5 shows the radial variations of energy flow intensities in the tube wall at various axial positions. The energy flow intensity should be larger at the position where the energy enters and get smaller along its travel in the tube wall because it is consumed as eddy current loss. This is the reason why the flow intensity become minimum in the tube wall. It is confirmed that the positions where the minimum values occur in Figure 4 coincide with the positions where the energy flows collide in Figure 4.

Figure 6 shows energy loss by eddy current in the tube wall. This figure also shows

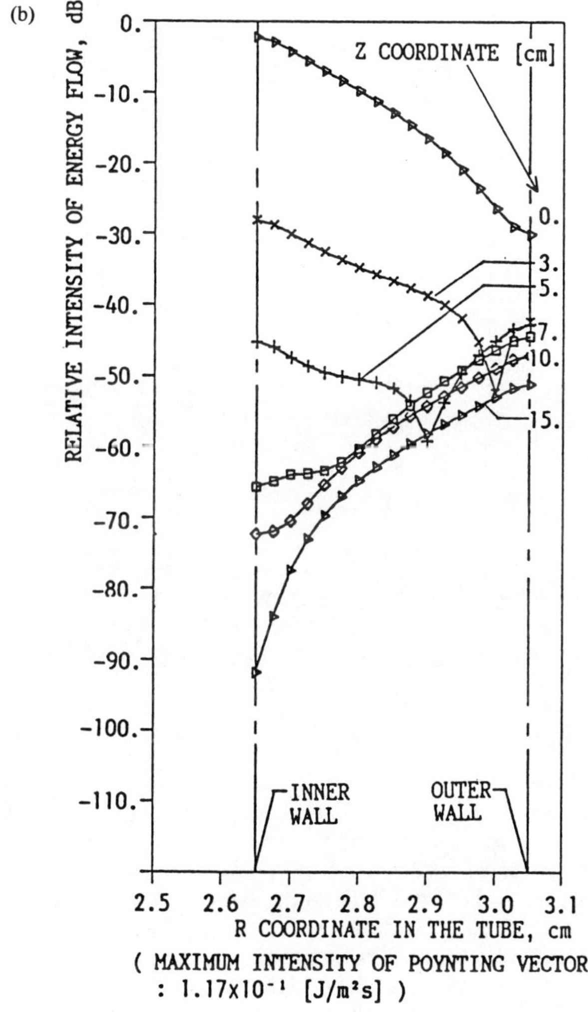

Figure 5 (b) Frequency: 120 Hz

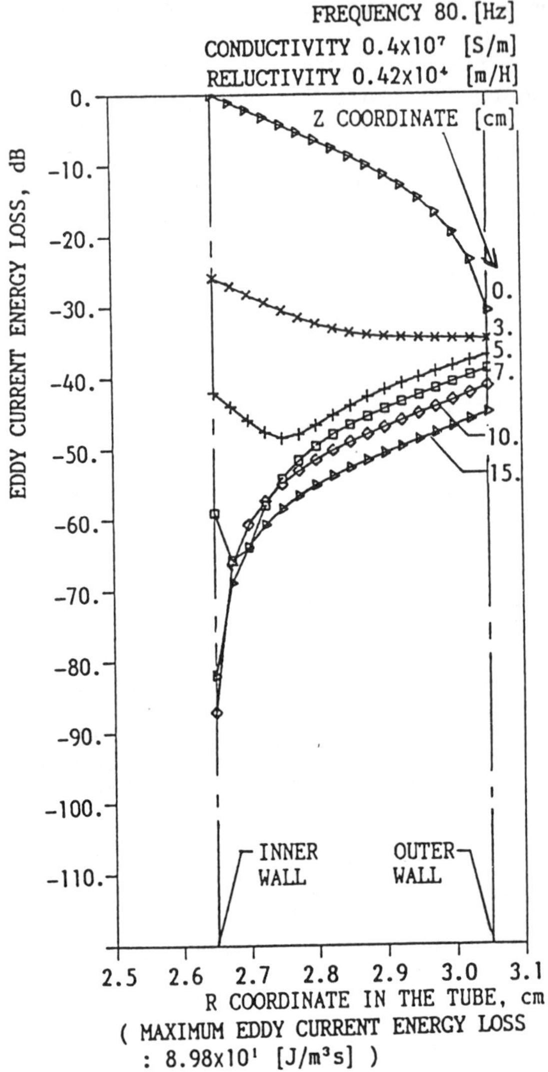

Figure 6   Eddy current losses in tube wall as functions of radial coordinate

an evidence of electromagnetic energy flow because the energy loss is considered larger where the intensity of energy flow is larger.

Figure 7 shows phase variations of vector potential at the same radial position with the exciter center as functions of axial distance from the exciter. These values mean the signal phase of detector coil. Comparing the results of Figures 2 and 4, the energy flow direction inside the tube is found to change in the area where the phase of vector potential leads. This fact is associated with the definition of Poynting vector by equation (1) because P depends on the phase difference between electric and magnetic fields.

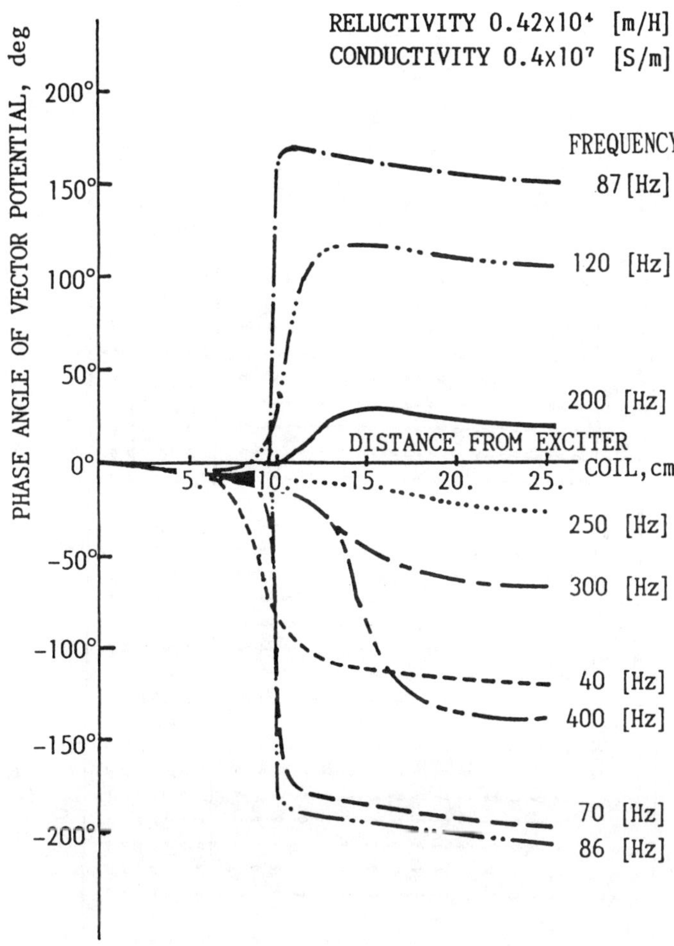

Figure 7 Phase of the vector potential inside and outside tube as functions of axial distance from the exciter

5. INFLUENCE OF DEFECT ON ENERGY FLOW

Figure 8 shows energy flow distributions when the tube has an encircling groove at outer surface in the RF area and in the NF area. In the figures only the radial coordinate in the tube wall is enlarged. It is shown in Figure 8(a) that the encircling groove in the RF area leads the flow from outside to inside causing distinct disturbance in energy flow distribution. Thus RFECT can detect defect in ferromagnetic tube because the detector picks up electromagnetic field disturbed by defect in the RF area. Figure 8(b) shows that an encircling groove axially located near the exciter can hardly lead energy flow from outside to inside of the tube and the energy flow distribution does not change inside the tube. Thus defect in ferromagnetic tube

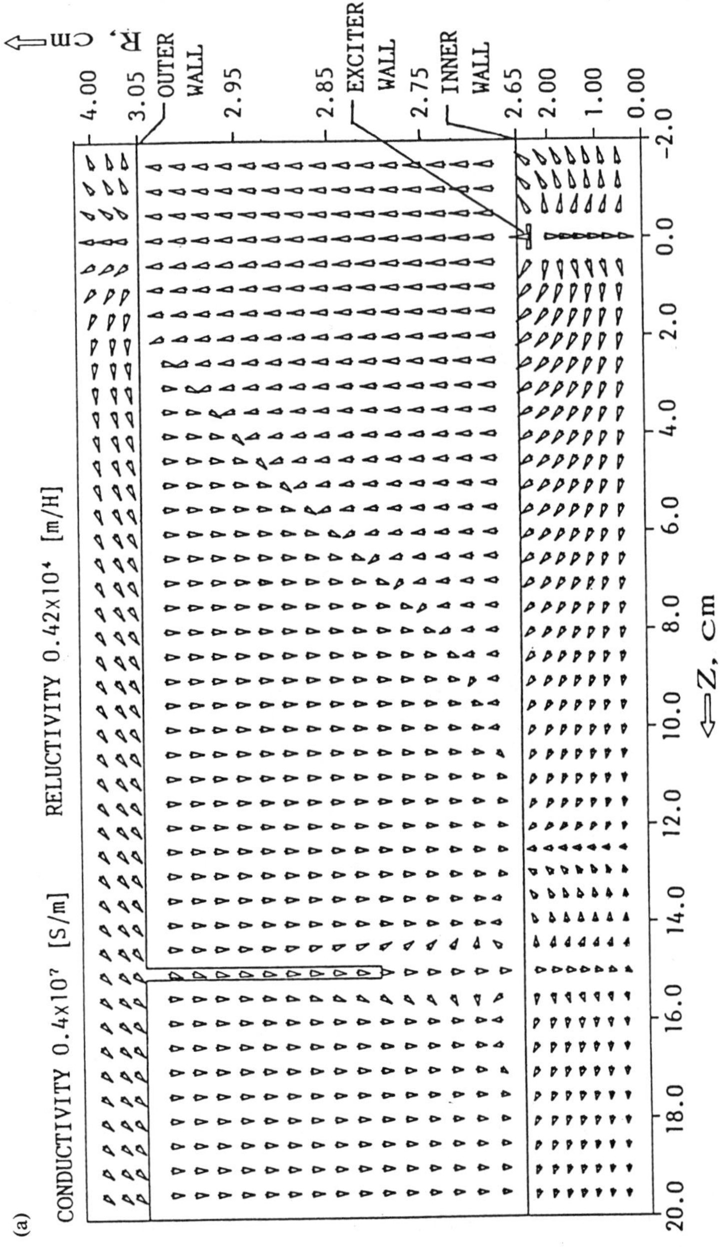

R, cm
cm
4.00
3.05
OUTER WALL
2.95
2.85
EXCITER WALL
2.75
INNER WALL
2.65
2.00
1.00
0.00
RELUCTIVITY 0.42x10⁴ [m/H]
CONDUCTIVITY 0.4x10⁷ [S/m]
Z, cm
-2.0
0.0
2.0
4.0
6.0
8.0
10.0
12.0
14.0
16.0
18.0
20.0
(a)

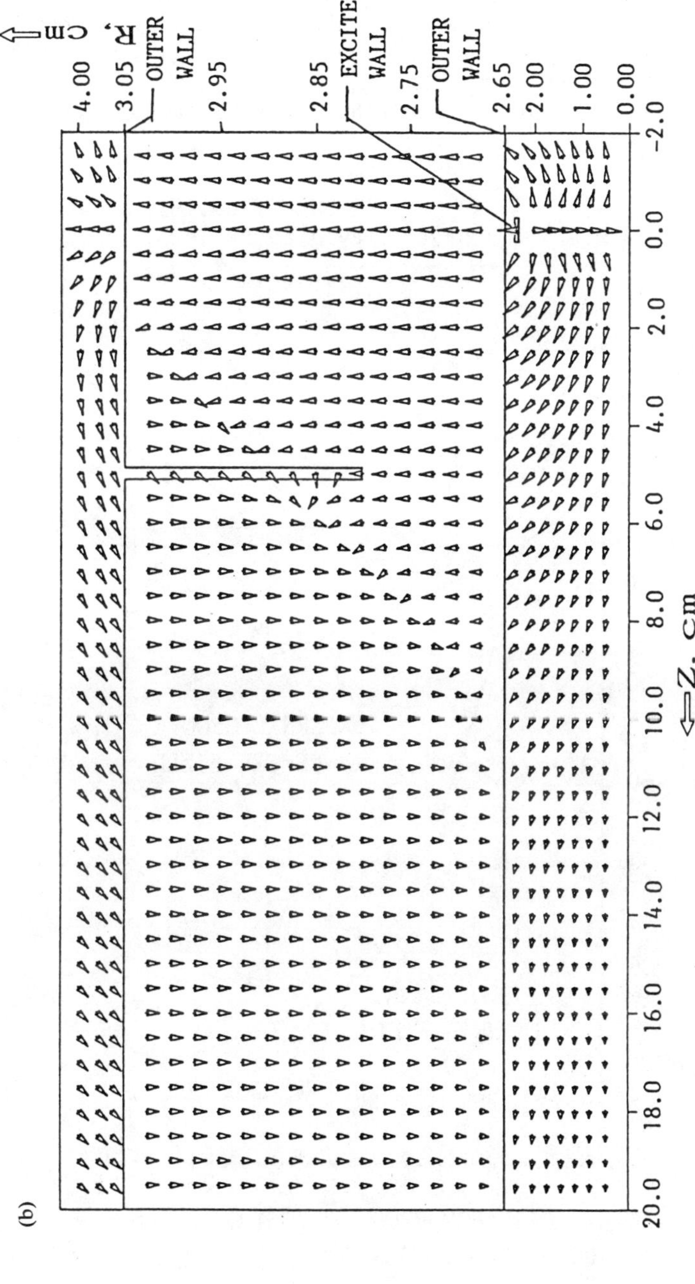

Figure 8 Variations of energy flow by an encircling groove at tube outer surface
(a) with a groove in RF area
(b) with a groove in NF area

can hardly be detected by the conventional eddy current testing which picks up the electromagnetic field disturbance in the NF area.

Figure 9 shows the variations of energy flow intensity at radial positions 2.5 mm apart from the inner and outer tube surfaces as functions of axial distance from the exciter when tube has an outer surface groove in RF area and in NF area. When a groove is in RF area, there is a distinct variation in the energy flow intensity. When a groove is in NF area, the variation of the intensity is very small.

Figure 10 shows the variations of energy flow intensities at the RF area at 15 cm

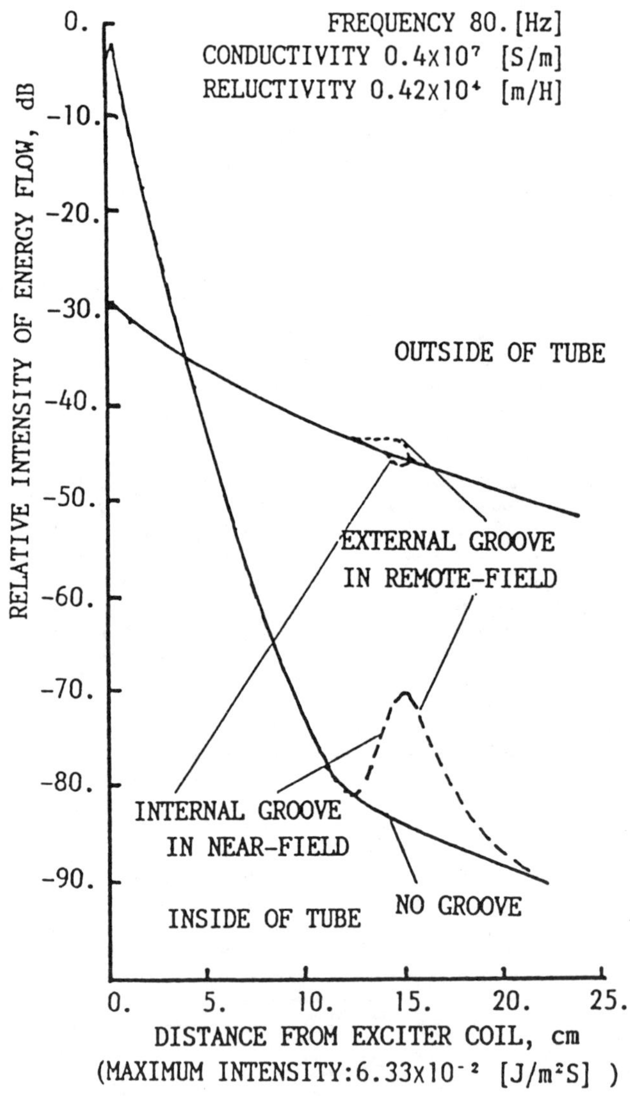

Figure 9 Energy flow intensities inside and outside tube with and without an groove

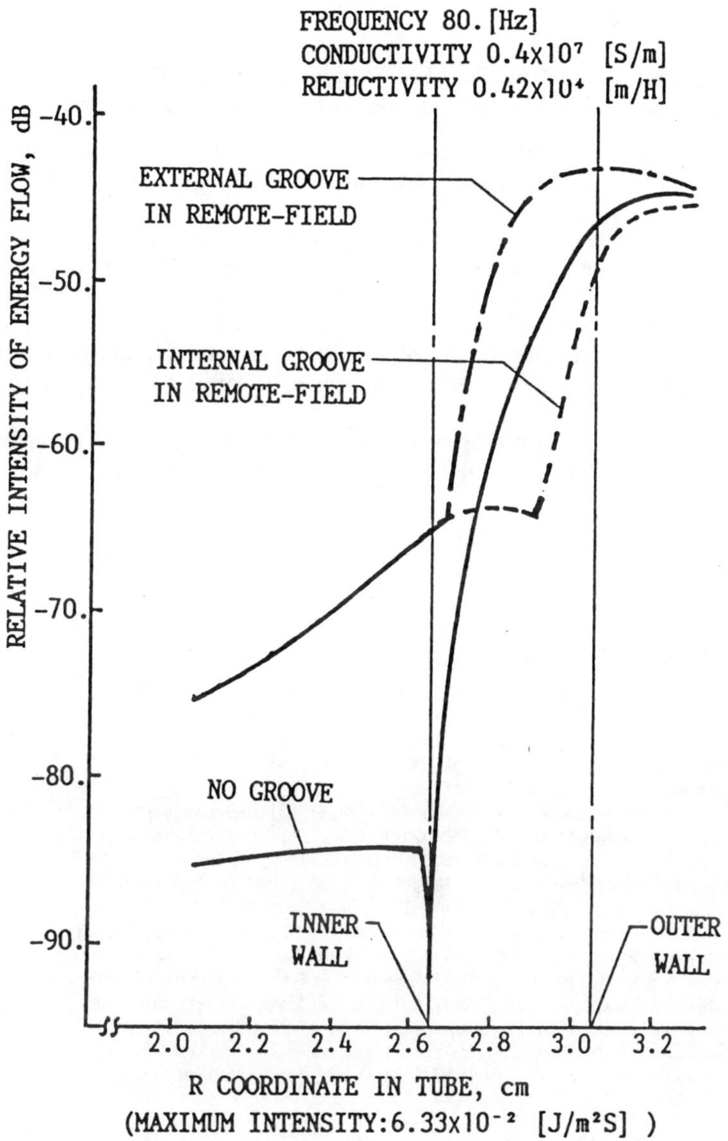

Figure 10 Energy flow variations in tube with and without a groove as functions of radial coordinate

displaced from the exciter. The solid line shows the variation for the tube without groove. The broken lines show for the tubes with a groove at inner and outer surfaces of tube. It is obvious from the figure that a groove decreases the decay rate of energy flow intensity in it. Besides, the flow intensities inside the tube are almost completely the same with inner and outer surface grooves. This fact supports that RFECT generates the same indications for inner and outer defects.

6. CONCLUSIONS

Electromagnetic energy flow has been analyzed quantitatively by finite element method to elucidate RFECT phenomena. The results are summarized as follows.

1. Inside the tube, the energy generated by the exciter is absorbed into tube inner wall and decreases very rapidly. As a result, the energy does not reach far from the exciter.

2. Some energy penetrating through the tube wall travels axially along the tube outer surface. The decay rate of energy flow outside the tube is so small that the intensity of energy flow becomes larger outside the tube than inside at the RF area.

3. Energy flows from inside to outside of tube in NF area and from outside to inside at RF area. Therefore, the detector in RF area can pick up the information on tube wall because RF area has energy penetrating through and carrying information on the tube wall.

4. When a groove is in NF area, the energy flow inside the tube hardly change. When a groove is in RF area, the flow inside the tube changes appreciably.

The authors believe that those results clarify phenomena underlying the remote field technique and will be useful for deciding the test conditions of the technique.

REFERENCES

1) T.R. Schmidt: *Corrosion*, **17**(7), 81–85, (1961).
2) S.G. Stroud, et al.: *J. of Petroleum Technology*, **14**, 257–260, (1962).
3) T.R. Schmidt: *Materials Evaluation*, **42**(2), 225–230, (1984).
4) D.L. Atherton, et al.: *Materials Evaluation*, **44**(13), 1544–1550, (1986).
5) D.L. Atherton, et al.: *British J. NDT*, **30**(1), 22–27, (1988).
6) K. Mori, et al.: *Japanese J NDI*, **37**(2A), 166–167, (1988) (in Japanese).
7) K. Mori, et al.: *Japanese J NDI*, **37**(2A), 168–169, (1988) (in Japanese).
8) Y. Hosohara, et al.: *Journal of JSNDI*, **39**(1), 9–12 (1990) (in Japanese).
9) Y. Sakamoto, et al.: *Journal of JSNDI*, **39**(1), 26–30, (1990) (in Japanese).
10) T. Kamimura, et al.: *Journal of JSNDI*, **39**(1), 38–43, (1990) (in Japanese).
11) H. Hoshikawa, et al.: *Journal of JSNDI*, **35**(2A), 160–161 (1986) (in Japanese).
12) H. Hoshikawa, et al.: *Journal of JSNDI*, **36**,(2A), 162–163, (1987) (in Japanese).
13) H. Hoshikawa, et al.: *Journal of JSNDI*, **36**(9A), 688–689, (1987) (in Japanese).
14) H. Hoshikawa, et al.: *Journal of JSNDI*, **37**(7), 568–579, (1988) (in Japanese).
15) H. Hoshikawa, et al.: *Materials Evaluation*, **47**(1), 93–97, (1989).
16) H. Hoshikawa, et al.: *Journal of JSNDI*, **37**(9A), 829–830, (1988) (in Japanese).
17) H. Hoshikawa, et al.: *Journal of JSNDI*, **38**(2), 185–186, (1989) (in Japanese).

ABSTRACTS OF PAPERS

Published in *Hihakaikensa* [*J. JSNDI*] **39** (1990)

No. 1

Energy Flow in Remote Field Eddy Current Testing

HIROSHI HOSHIKAWA, HIDEYUKI SAITOU, JUNJI KOIDO and YASUO ISHIBASHI

College of Industrial Technology, Nihon University, 1-2-1 Izumi-chō, Narashino-shi, Chiba 275, Japan

Remote field eddy current technique (RFECT) has been considered as a promising in-service inspection method for steel tubing. However, there are many phenomena to be clarified such as why defects are detected through the method. The authors have investigated the energy flow in remote field to elucidate the phenomena.

The electromagnetic energy generated by the exciter coil decays rapidly inside the tube, and the energy outside the tube travels with small decay rate along the tube. As a result, the energy flow intensity outside the tube is larger than that inside the tube and the remote field is dominated by the energy from the outside. Consequently, RFECT can detect defects in remote field area because there exists the energy which contains the information on the tube wall.

(page 31)

No. 2

Experimental Studies on Detectability of Remote Field Eddy Current Testing Method

TAKEO KAMIMURA

Mitsubishi Heavy Industries, Ltd. Takasago R & D Center, 2-1-1 Arainiihama, Takasago-shi, Hyōgo 676, Japan

For the purpose of comprehending the effect in practical use of the Remote Field Eddy Current (RFEC) Testing that becomes noticeable for the ISI technique of steel tubes, its flaw detectability was clarified through a model test.

This study used straight and bending tubes of 3.8 mm in wall thickness and 31.8 mm in outside diameter. These tubes were inspected from their inside. After relations among the pickup coil output, coil distance, testing frequency, etc. were measured, a probe of the practical use type was manufactured on trial to investigate its flaw detectability by means of simulated flaws.

Consequently, it has been found that light local wall thinning on the outside surfaces can be detected by this technique and its effect in practical use can be expected with small influences due to magnetic permeability variations of tube materials, bending of tubes, etc.

(page 38)

No. 2

A New X-Ray Technique for Evaluating Hardening of Quenched Steels

MASANORI KURITA, MASASHI SHINDŌ*, MATSUO MIYAGAWA

*Nagaoka University of Technology, 1603–1 Kamitomikoa-chō, Nagaoka-shi,
Niigata 940–21, Japan*
** Niigata Sanyō Electric Co., 3000 Chiyakō, Ojiya-shi, Niigata 947, Japan*

An X-ray diffraction line profile will broaden remarkably when steels transform to a martensitic structure on quenching. The profiles for steels having a ferritic structure will sharpen. Slackly quenched steels, on the other hand, have an anomalous diffraction profile with a sharp peak superimposed on a broad base because they have a mixture structure of martensite and fine pearlite having ferritic structure. A new technique is proposed for evaluating the extent of hardening of quenched steels from the broadness of the diffraction peak represented by the standard deviation of a Gaussian function fitted to the diffraction peak. This standard deviation is called Gaussian curve parameter (GCP). By using fifteen types of steel of carbon contents between 0.16 and 1.45%, the GCPs of quenched steels were measured. The GCPs of slacky quenched steels fall between those of completely hardened steels and annealed steels. The comparison of the GCP of the specimen to that of the completely hardened steel permits rapid and nondestructive evaluation of hardening of quenched steels without measuring the hardness.

(page 67)

Propagation of Ultrasonic Wave in Anisotropic Materials

SADAO HIROSE, HIROSHI URAGAKI, YUJI MATSUMOTO*,
OSAMU KOBAYASHI* and YOSHIKAZU NISHIMOTO**

Ōsaka University, Faculty of Engineering, 2–1 Yamadaoka, Suita-shi, Ōsaka, Japan
** Nissho Iwai Co., Ltd., 3–30 Imahashi, Chuō-ku, Ōsaka 541, Japan*
*** Shin-Nippon Nondestructive Inspection Co., Ltd., 4–10–13 Ihori, Kokurakita-ku,
Kitakyūshū-shi, Fukuoka 803, Japan*

The propagation characteristics of ultrasonic wave were examined in anisotropic materials. A control rolled steel plate and an aluminium alloy plate were used to measure the velocities of transverse wave propagating in various directions. On the results of measurements, a relation between propagating directions and velocities was obtained as a linear combination of cosine functions.

Applying this expression, propagating directions of ultrasonic waves in anisotropic materials were analyzed based upon Huygens' principle. As a result, it was obtained an expression of relation between the propagation angles in anisotropic materials and the refraction angles in the type A1 calibration block. The propagation angles measured in the rolled direction and the cross direction of control rolled steel plate provided a good correspondence with calculated results.

Furthermore supposing various distribution of wave velocities', the influence of anistropy upon the propagating direction was evaluated.

(page 74)

An Experimental Study for Beam Displacements of Corner Reflection in Ultrasonic Angle Beam Testing

MASAKAZU TAKAHASHI

Nippon Chūtetsukan KK, Kuki Shōbu Works, 1 Shōwanuma, Shōbu-machi, Minamisaitama-gun, Saitama 346–01, Japan

An experimental study for beam displacements of corner reflection in ultrasonic angle beam testing was carried out with the use of variable angle beam probes. In total reflection condition of both surfaces, a right angled corner in steel specimen, a probe distance varies from position of geometrical calculation by beam displacement, on that reason the beam path length varies. Apart from 45 degrees in beam angle the probe distance is shifted backward in 33.2° to 44° and beam path length becomes longer, the probe distance is shifted forward in 46° to 59° the beam path length becomes shorter, the shift can be explained well with Artman's theory for total reflection condition. In beam angel 60° to 70° the shift is not so large in comparison with the above because SV wave reflection from one of two surfaces is not in total reflection, but is affected by reflection coefficient in opposite side to beam displacement. We should take those shifts into consideration in the evaluation of defect location and height in ultrasonic angle beam testing.

(page 80)

Experimental Procedure to Evaluate Impact Tensile Strength of Brittle Polymer Material

HIROKI TODA, HIDEKAZU FUKUOKA, FUMIO INOUE, TATSUAKI SAWAI and YUKIO UEMURA*

Ōsaka University, 1–1 Machikaneyama-machi, Toyonaka-shi, Ōsaka 560, Japan
** Hannan University, 5 Amamihigashi, Matsubara-shi, Ōsaka 580, Japan*

It is well known that it is very difficult to evaluate the impact strength of brittle material by making use of the usual test machine such as Charpy impact test with force detection system, because the flexural vibration induced by a lateral impact disturbs estimation of the stress in the test specimen.

In this paper, we proposed an experimental procedure of tensile impact test to evaluate the strength of brittle polymer material under high loading rate.

We have applied our method to the measurement of dynamic fracture toughness of polymethyl-metacrylate plate specimen with sharp notches by using the trial equipment, which was reformed from the Charpy test equipment. We have got the validity of this method, and also have obtained new knowledge about the dependence of dynamic fracture toughness of PMMA on the loading rate.

(page 86)

No. 3

Measurement of Density by Back Scattered X-rays

MAMORU MIZUNUMA

Nippon Industrial Technology Material Testing Center, 4–1 Gakuendai, Miyashiro-chō, Minamisaitama-gun, Saitama 345, Japan

Density-measurement by using back scattering of X-rays instead of gamma-rays have been studied. The back scattering were measured at 2 points of surface length of scatterer for eliminated the bad effect be caused by unstable X-rays.

The difference in distance of 2 points having constant ratio of the scattering and the ratio of scattering at 2 points having constant difference in distance were measured, and studied the relationship between the density and the ratio of scattering or the difference in distance of 2 points in theory and experiment.

Furthermore, the relationship between the several unstable factors and accuracy of measurement were studied. These factors were the thickness of scatterer, the fluctuation of X-rays, and the setting error of incident angle. The results were shown that the thickness needs some dimension according to the tube voltage, and that the higher tube voltage and the larger exposure have better accuracy, and that at high incident angle, these measurements are little affected by the setting angle error.

(page 211)

Advanced Technique and Instrument to Measure Thickness of Concrete

TETSUO YAMAGUCHI

Tōyoko Elmes Co., Ltd, 2-18-2 Minamimaioka, Totsuka-ku, Yokohama-shi, Kanagawa 244, Japan

An advanced ultrasonic wave technique to measure the thickness of thick concrete blocks from one sided surface, was designed and manufactured. The measuring principle was based on the separation of a reflected ultrasonic wave, from ultrasonic vibration waves at the surface of concrete near the probes. Fourier analysis by FFT analyzer was used for the purpose. Measured results on the thickness of concrete block of 200 mm and 500 mm thick respectively, and a cavity detection in a concrete block of 750 mm thick were shown. Many field tests in tunnels and floors were carried out, and some of experimental phenomena were also presented in this paper.

Ultrasonic Attenuation on Various Materials under Elastic Stress

KENICHI KATSUMATA

Ship Research Institute, 6-38-1 Shinkawa, Mitaka-shi, Tokyo 181, Japan

Ultrasonic waves were applied to various materials under stress where the ultrasonic wave propagation direction was perpendicular to the stress direction. Materials studied were steel, stainless-stell, aluminium, copper and acrylic-resin. The attenuation was measured also on steel specimens with applied magnetic field.

It was found that the ultrasonic attenuation and wave direction were sensitive to both the tension and compression stresses in each material. Steel clearly exhibited attenuation hysteresis but for the materials under elastic stress, hysteresis was low or non-existent. It is suggested that in steel vibration losses occur because of the magnetic domain by the applied stress. The relevance of these results to the nondestructive evaluation of materials properties is discussed.

(page 223)

Non-destructive Evaluation of Porosity in Sintered Iron by Magnetomechanical Acoustic Emission

NOBORU SHINKE, YOSHITUGU OHIGASHI*, AKIKAZU NAKAGIRI,
TAKAAKI YAMANO

*Faculty of Engineering, Kansai University, 3-3-35 Yamate-chō, Suita-shi, Ōsaka 565, Japan
* Ōsaka Titanium Co., Ltd., Japan*

The discontinuous motion of magnetic domain walls in ferromagnetic materials during magnetization produces magnetomechanical acoustic emission. The magnetomechanical acoustic emission has characteristics which show a sensitive dependence on metallographical structure of ferromagnetic materials.

In this paper, the possibility of non-destructive evaluation of porosity in sintered iron has been mainly investigated by using the magnetomechanical acoustic emission. The results obtained are summerized as follows: (1) The higher porosity in sintered iron increases the magnetomechanical acoustic emission intensity, either AE count or amplitude. (2) Evaluating the effects of porosity with the values normalized by magnetomechanical acoustic emission intensity in bulk material, it is recognized that there is more remarkable porosity effect on mean event energy of magnetomechanical acoustic emission. (3) The results of measurements suggest an adequate correlation between the porosity and the magnetomechanical acoustic emission in sintered iron.

(page 231)

Effects of WC Grain-size and Co Contents on Fracture Toughness of WC-Co Alloys

KENKICHI SATŌ and HARUHIKO HONDA*

*Department of Mechanical Engineering, Chiba University, 1-33 Yayoi-chō, Chiba 260,
Japan
* Development Div., Hitachi Tool Eng., Japan*

The fracture toughness K_{IC} of WC-Co alloys was measured by the method of four-point-bending with chevron-notched specimens (CN-method) and by the method of Vickers indentation fracture (IF-method). The alloys used had WC grain size d_{WC} in the range of 0.6 to 6.0 μm and Co content f_{CO} in the range of 5 to 20 wt%. The value of K_{IC} measured by the CN-method was 10 to 24 MPam$^{1/2}$ and the value measured by the IF-method was 30 to 80% higher than that by the CN-method for the low hardness alloys. As the result, the application of the IF-method was limited to the higher hardness alloys. The value of K_{IC} increased with increasing WC grain size and Co content. This dependence for the result of CN-method was generalized by using a parameter of mean free path of cracks in Co phases. A formula of the relationship between K_{IC} and Hv (Vickers hardness) was also obtained experimentally for the alloys used.

(page 237)

A Basic Theory for X-Ray Stress Measurement by Gaussian Curve Method

MASANORI KURITA

*Nagaoka University of Technology, 1603-1 Kamitomioka-chō, Nagaoka-shi,
Niigata 940-21, Japan*

In X-ray stress measurement, the stress value is calculated from several peak positions of the diffraction lines from lattice planes differently oriented with respect to the specimen normal. In the Gaussian curve method, the peak position of the diffraction line is determined as the main axis of a Gaussian function fitted to n data points around the diffraction peak by using the least squares method. This equation in its original form is very complicated. A simplified equation for calculating the peak position is derived for rapid calculation and to avoid the error due to the overflow of figures in the calculation with a computer. In the half-width method, which is most widely used in Japan, the background subtraction is needed in determining a peak position, but it is unnecessary in the Gaussian curve method. The Gaussian curve method can determine even a peak position of an extremely broad diffraction line of a hardened steel and those of diffraction lines being located closely, for these diffraction lines, the half-width method cannot be used because the background lines are undeterminable.

(page 243)

Development of Truck Self Weighing System by Strain on the Axle of Truck

YOSHIROU NOJIRI

Kyowa Electronic Instruments Co., Ltd., 3-5-1 Chōfugaoka, Chōfu-shi, Tōkyō 182, Japan

There is a way to measure cargo weight, using the system which is installed on the instrument panel of driver's cabin.

As one of the ways, we found that the strain, which is on the side part of the differential housing case of drive axle, is in proportion to a cargo weight, so we studied to develop for applying it to the practical system.

In the first place, we tested the accuracy of the measurement and its environment durability, using the sensor, with easily installing and reliability for long time, which we could install to the axle.

By static load test we found that the accuracy of the measurement was $\pm 5\%$.

Next time, we improved the indicator to be used easily by average drivers. Using this improved system, we measured the actual cargo weight for running 100,000 km. As a result, we confirmed that the actual accuracy of the measurement was $\pm 10\%$.

(page 250)

No. 4

Considerations on the Sensitivity Difference between the Inside and Outside Defects in the Ultrasonic Angle Beam Testing for Pipe and Tube

MASAKAZU TAKAHASHI

*Nippon Chūtetsukan KK, Kuki-Shōbu Works, 1 Shōwanuma, Shōbu-machi,
Minamisaitama-gun, Saitama 346-01, Japan*

Experimental studies were carried out for sensitivity difference between the inside and outside defects in the angle beam testing for pipe and tube. Echo heights of the inside and outside defects of 0.30 mm depth were measured at 34.0° to 43.6° refracted angles in 50 mm OD, 4 to 9 mm wall thickness pipe. The reason for sensitivity difference between the outside and inside defects seems to be the divergence of beam reflection of defect, the loss of mode conversion on inside defect surface, the reflection directivity of defect and so on. The experimental result was not explained by Werneyer and Ammirato theories based on the reflection directivity of defect. Hence other considerations are required.

(page 307)

Measurement of Fatigue Crack Closure in Polymer by Using Photoelastic Technique. First Report: On the Change of Mechanical and Photoelastic-plastic Effect Associated with Cyclically Plastic Deformation in Polycarbonate

SHUNJI NAGASAKA

Kōgakuin University, 1-24 Nishishinjuku, Shinjuku-ku, Tōkyō 160, Japan

As one of the basic research to measure the fatigue crack closure and the effective stress range for crack opening in polymers by using photoelastic technique, the change of photoelastic properties and strain accumulation caused by cyclically plastic deformation in polycarbonate plain specimens were investigated. Although the relationship between principal strain and photoelasto plastic effect at plastically deformed region around crack tip was shown as a bilinear relation in the region of small cycles, its relation came to change into linear with the increase of cyclic numbers especially at the highly strained plastic zone of polycarbonate.

(page 316)

Study of Ultrasonic Attenuation of Steel Materials in a Magnetic Field
Part 1: Effects of Propagation Distance and Direction

KENICHI KATSUMATA

Ship Research Institute, 6-38-1 Shinkawa, Mitaka-shi, Tōkyō 181, Japan

The effect of a change in magnetization on the ultrasonic attenuation of 40–50 kg/mm^2 tensile strength specimens, using longitudinal and shear waves was studied. The difference in the attenuation of the received amplitude of magnetized and non-magnetized steel was defined as the Magnetization Effect Attenuation Sensitivity (MEAS). Integrated MEAS was defined as U. Results were as follows:

(1) MEAS increased proportionally with ultrasonic propagation distance when the longitudinal wave direction was perpendicular to the magnetic field.

(2) When the wave vibration direction was perpendicular to the magnetic field and normal shear wave was applied, MEAS and U values were larger than when the wave vibration direction was parallel to the magnetic field.

(page 323)

No. 5

Study of Ultrasonic Attenuation of Steel Materials in a Magnetic Field
Part 2: Effects of Elastic Stresses and Pre-Strain

KENICHI KATSUMATA

Ship Research Institute, 6-38-1 Shinkawa, Mitaka-shi, Tōkyō 181, Japan

The Magnetization Effect Attenuation Sensitivity (MEAS) was measured under elastic stresses in 40 (SM41B) and 50 kg/mm^2 (SM50A) tensile strength steel specimens at magnetic field strength of 0-330 Oe. MEAS was positive and decreased with increasing applied tensile stress. But with increasing compressive stress, MEAS increased in the positive domain for a shear wave vibration direction perpendicular to the stress direction. It decreased to become negative for a shear wave vibration direction parallel to the stress direction.

Pre-strain of about 1% and 3% produced negative MEAS values with increasing applied magnetic field. However, for a shear wave with vibration direction perpendicular to the applied stress, MEAS values slowly increased to become positive for both pre-strains at magnetic field of 60 and 130 Oe (SM41B).

(page 369)

Development of the High Sensitivity and Low Noise Integrated AE Sensor

MITSUHARU SHIWA, HIDEHIRO INABA*, TERUO KISHI

*Research Center for Advanced Science and Technology, University of Tōkyō, 4-6-1
Komaba, Meguro-ku, Tōkyō 153, Japan
* Fuji Ceramics Co., Ltd., Japan*

High sensitivity and low noise acoustic emission sensor is developed. The response of the sensor is analyzed both from the response behavior of piezoelectric device (PZT) and network signal loss of the circuit. The resonance mode of radial and thickness vibrations in circular rod PZT is studied. Therefore, the highest sensitivity is obtained to set the same frequency resonance both the radial and the thickness mode.

Developed equivalent circuit model is analyzed for the signal loss and the noise as the assembly of PZT, a signal cable and a pre-amplifier. From this analysis, three factors are obtained for a good design of a circuit: a high capacitance PZT, direct coupling PZT and pre-amplifier and low capacitance input of pre-amplifier. Then, the integrated AE sensor with complete shield-casing is developed using the direct coupling between the selected PZT and FET.

This sensor shows low noise level as following: 14 μV (100 kHz to 1.2 MHz) and 24 μV (100 kHz to 20 MHz) at the first stage of the circuit: 1/2 lower signal loss compared with normal system case.

(page 374)

Measurement of Fatigue Crack Closure in Polymer by Using Photoelastic Technique. Second Report: Variation of Effective Crack-Opening Ratio in Crack Closure in Polycarbonate

SHUNJI NAGASAKA

Kogakuin University, 1-24 Nishishinjuku, Shinjuku-ku, Tōkyō 160, Japan

Fatigue crack closure in polycarbonate polymeric material was measured by using photoelastic technique. From hooked point on the relationship between applied stress (σ) and photoelastic fringe number (N) at the crack tip, fatigue crack opening stress (σ_c) and effective crack opening ratio Up ($= (\sigma_{max} - \sigma_c)/\sigma_{max}$) were determined. Relationship between Up and stress ratio R ($= \sigma_{min}/\sigma_{max}$) increased linearly in the range of $R \geqq 0.2$ as similar as Elbers experimental formula, however, it was shown as a different curve from Elbers formula in the range of $0 \leqq R \leqq 0.2$. The validity of the estimation based on Up approach was also confirmed in the case of crack growth retardation caused by a spike overload.

(page 383)

Measurement of Impact Load in Instrumented Impact Testing

HIROTSUGU INOUE, TOSHIKAZU SHIBUYA, TAKASHI KOIZUMI and
KIKUO KISHIMOTO

Tōkyō Institute of Technology, 2-12-1 Ōokayama, Meguro-ku, Tōkyō 152, Japan

An instrumented impact testing is widely used for evaluating mechanical properties of materials subjected to impact loading. In such testing, an impact load applied to specimen is one of the most important data. However, there are many difficulties in measuring the impact load accurately. Most of the previous methods ignore the fact that the impact load propagates in an impactor as a stress wave. Therefore, it can be said that they have some inaccuracy. Our study is concerned with a method to evaluate the impact load by applying an inverse analysis to the impact response of the impactor. The wave propagation in the impactor is taken into account in our method. In this paper, we applied this method to the instrumented Charpy impact testing, and discussed about the measurement of impact load in instrumented impact testing. As a result, it is shown that the impact load is measured successfully by our method, and problems which are associated with the previous methods are pointed out.

(page 390)

No. 6 Annual Reviews of Technical Committees, and no Papers.

No. 7

Evaluation Technique for Metal Properties by Analyzing Ultrasonic Spectrum

KATSUHIKO HONJOH, YOSHIKAZU SUDOH* and JUN-ICHI MASUDA

NTT Technical Assistance & Support Center, 3-9-11 Midori-chō, Musashino-shi, Tōkyō 180, Japan
** AIREC Engineering Co.*

The purpose of this study is to clarify quantitatively the relationship between ultrasonic pulse spectrum pattern and mechanical properties in metals. Analyzing ultrasonic pulse echo waveforms in the frequency region, which were extracted in the time region equivalent to one or two wavelength, so that the results were obtained as follows, 1) Peak frequency of ultrasonic pulse echoes is not varied in near field and is reduced in far field as propagation length increased. 2) Peak frequency of ultrasonic pulse echoes increases linearly as ultrasonic propagation velocity of material increases. 3) Variation in the pulse peak frequency as a function of material velocity of propagation length increases as increment of stimulated frequency for the transducer. In addition, we have discussed these results adopting attenuation theory due to the scattering and the spatial diffusion of ultrasonic pulses propagating in metals.

(page 587)

Study of Ultrasonic Attenuation of Steel Materials in a Magnetic Field
Part 3: Effects of Specimen Size

KENICHI KATSUMATA

Ship Research Institute, Shinkawa, Mitaka-shi, Tōkyō 181, Japan

The ultrasonic attenuation was measured within the magnetic field in steel plates. Effects of test specimen size were examined 40 kg/mm^2 (SS41) tensile strength steel plate 12 mm thick.

It is demonstrated that the Magnetization Effect Attenuation Sensitivity (MEAS) is decreased by increasing specimen width and becomes constant for specimens 4 times wider than the magnet. For these specimens MEAS value decreased to 1/4 of the specimen with the same width as the magnet. These results can be applied to nondestructive evaluation of the change in material properties for the structural steel parts.

(page 594)

Relationship between Fatigue Crack Extension Curve and Striated Pattern of Fracture by Optical Microscope Fractographic Study of Fatigue Fracture Surface of Polycarbonate.
1st Report

SHUNJI NAGASAKA

Kōgakuin University, 1–24 Nishishinjuku, Shinjuku-ku, Tōkyō 160, Japan

Fractographic study with an optical microscope for a fatigue fracture surface of polycarbonate is presented. Polycarbonate shows discontinuous crack growth behaviour under cyclic loading: the fatigue crack grows every length of a plastic zone constructed at the fatigue crack tip. The relationship between a striated pattern observed at the fatigue surface and a characteristic fatigue crack extension curve is also observed.

(page 607)

No. 8

Propagation Characteristic of Stress Pulse for Mg Produced Heated Mold Continuous Casting Method

HIROYUKU TAKEISHI, TAKESHI KUDŌ, TOMOYA MINEGISHI and ATSUMI OHNO

Chiba Institute of Technology, 2–17–1 Tsudanuma, Narashino-shi, Chiba 275, Japan

Pure Mg plate casted by heated mold continuous casting method (O.C.C.-Mg) have layer structure of the columner crystals parallel to the casting direction. The mechanical and acoustic properties of O.C.C.-Mg is compared with that of pure Mg casted by conventional casting method (polycrystal Mg). One dimensional stress wave is used for investigation of their properties. The gas-gun is used for generation of stress wave and strain gages are used for detection of stress wave.

Young's modulus, elastic limit and Poisson's ratio for the stress wave are inquired. The amplitude and duration time of the stress wave dependence of the attenuation constant are confirmed.

Moreover, frequency distribution which is connected to attenuation of stress wave in O.C.C. or polycrystal Mg is conjectured by Fourier analysis.

(page 642)

Measurement of Contact Load by the Method of Caustics

KENKICHI SATŌ, MUNENOBU SATŌ, HIROAKI ITOGA and YASUO NAMAIZAWA

Faculty of Engineering, Chiba University, 1–33 Yayoi-chō, Chiba 260, Japan

The method of caustics was applied to determine the contact load in the case of Hertzian contact between a plate and a disk, and then the limitation of applicability of the method was defined. The experiments were performed with the plate of PMMA and disks of PMMA and steel whose radii are in the range from 1 to 50 mm. The contact load was determined by measuring the maximum diameter of the caustic image on the screen. The contact load determined by this method was valid for the case of the caustic images which had the value of Y/X between 0.94 and 1.02, where X and Y denote the size in the x and y directions of the caustic images, respectively. This limitation is due to the difference between the concentrated load in the theory and the distributed load in the practice. In order to expand this limitation of the method of caustics, a modified theory for the distributed load as the Hertzian contact was proposed. By the modification the method of caustics is applicable to the contact between the plate and the disk of the radius of 25 mm.

(page 648)

Local Strain Behavior at the Notch-root and the Crack-tip on Corrosion Fatigue

TADASHI SATŌ, HEIHACHI SHIMADA*, KANEAKI KIMURA and
HIROYUKI IGUCHI**

*Department of Mechanical Engineering, Faculty of Engineering, Iwate University,
4-3-5 Ueda, Morioka-shi, Iwate 020, Japan*
** Akita National College of Technology, 1-5 Iijimabunkyō-chō, Akita-shi, Akita 011,
Japan*
*** Toyota Motor Corporation, 1 Toyota, Toyota-shi, Aichi 471, Japan*

Much attention has been given to corrosion fatigue because many structures and machine components operate in corrosive environments. Many studies on corrosion fatigue in various environments have been completed. However, these studies did not deal with local strain behaviour near the notch-root and the crack-tip in corrosive environments, because local strain measurement method in high temperature and corrosive environments has not yet been established.

In this paper, local strain at the notch-root and crack-tip was investigated by a fine-grid method for corrosion fatigue of Type 304 stainless steel in boiling 42% $MgCl_2$ solution.

On the basis of strain histories at notch-root and crack-tip, the strain range under cyclic loading conditions during each cycle was investigated in detail.

The correlation between local strain range and number of cycles to crack initiation at notch-root is proposed. This approach differs fundamentally from the usual method proposed by Morrow, J. and the others.

Moreover, the relationship between the crack propagation rate to local strain range at crack-tip can be expressed linearly on the wide range of crack propagation rate.

(pg 654)

A Method for Determination of X-ray Elastic Constants of Materials Showing Nonlinear $\sin^2\psi$ Diagrams and Its Application to Zn-Ni-alloy Electroplate

TOSHIHIKO SASAKI, MAKOTO KURAMOTO and YASUO YOSHIOKA*

*The Institute of Vocational Training, 1960 Aihara, Sagamihara-shi, Kanagawa 229, Japan
* Musashi Institute of Technology, 1-28-1 Tamatsutsumi, Setagaya-ku, Tōkyō 158,
Japan*

This paper describes the method and the experiment for the determination of the X-ray elastic constants of Zn-Ni-alloy electroplate. For this material, the $\sin^2\psi$ method is not adequate to use because this material shows severely curved $\sin^2\psi$ diagrams. Therefore, a new method developed by the authors was explained first. This new method is effective for materials showing nonlinear $\sin^2\psi$ diagrams.

Secondly, the experiment was made on the application of this method to the Zn-Ni-alloy electroplate. And it was found out that the experimental data agreed well to the theory of this method.

As a result, the following values were obtained as the x-ray elastic constants of the sample measured:

$$\frac{(1+\nu)}{E} = 8.44\,\mathrm{TPa}^{-1}\,\frac{\nu}{E} = 2.02\,\mathrm{TPa}^{-1}.$$

(page 660)

Application of Imaging Plate to Micro-beam X-ray Diffraction

YASUO YOSHIOKA, SHIN-ICHI OHYA and TAKESHI SHINKAI

Musashino Institute of Technology, 1–28 Tamatsutsumi, Setagaya-ku, Tōkyō 158, Japan

A new type of integrating area detector system with high sensitivity and high spatial resolution was recently developed for diagnostic radiography. In this detector system, a two dimensional X-ray image is temporarily stored as a distribution of F-centers in a photostimulable phosphor screen called the imaging plate (IP). The image in the IP is then read out by measuring the intensity of fluorescence which is stimulated by a focused He-Ne laser beam scanning the surface of the phoshor screen. The residual X-ray image in the IP can be erased simply by exposing it to a large dose of visible light and the IP can be used repeatedly. The detector has 100% detective quantum efficiency for 0–20 keV X-ray, a spatial resolution better than 0.15 mm (fwhm), a dynamic range of 10^5 and no counting rate limitation. The exposure time can be shorten to 1/20–1/60 in comparison with the use of the X-ray film.

In this study, we examined the possibility of the IP for the X-ray studies on the mechanical behaviour of materials by using the back-reflection X-ray technique. An exposure time of more than 30 minutes would be required for a conventional high sensitivity X-ray film in the case of α Fe (211) diffraction by Cr-Kα X-rays. When the imaging plates were used in place of the film under the same X-ray condition, we could obtain visually similar patterns by exposing the time of less than 90 seconds. These diffraction patterns can be precisely analyzed with the help of the image processing analyzer. We conclude that this detector system is usable in almost the same way as an X-ray film. Especially, this will be more powerful means in the field of micro-beam X-ray diffraction.

(page 666)

Dependence of Barkhausen Noise Signals on Sensor and Its Calibration Method

NORIHIKO NAKAI and MITSUO OBATA*

*Nippon-Kosuha Steel Co., Ltd., 1-7-2 Ōtemachi, Chiyoda-ku, Tōkyō 100, Japan
* Department of Materials Processing, Faculty of Engineering, Tōhoku University,
Aramaki-aza Aoba, Aoba-ku, Sendai-shi, Miyagi 980, Japan*

The purpose of this paper is to develop a method to calibrate the sensor ("Yoke" type sensor) dependence of Barkhausen noise (BHN) signals (V_p). For this purpose, first the authors theoretically investigated sensor dependence of V_p by applying a magnetic circuit theory. The results showed that there is a proportional relation between V_p's calculated with various conditions. Then, this relation was experimentally confirmed by using four different sensors which had been prepared for this study. Summarizing theoretical and experimental results, the authors propose a method to calibrate the "Yoke" type BHN sensor dependence of of of V_p.

(page 672)

Development of Noncontact Shape-Measurement by Applying Digital Image Correlation

TAKASHI FURUKAWA and MITSUO OBATA*

*Department of Materials Processing, Faculty of Engineering, Tōhoku University,
Aramaki-aza Aoba, Aoba-ku, Sendai-shi, Miyagi 980, Japan*

One of the authors developed and reported a digital image correlation method to measure two-dimensional surface displacements. The method has the advantage of measuring deformations on the rough surface without contact. This means that a hybrid method of stereoscopy and digital image correlation technique is hopeful of three dimensional shape measurement without surface polishing. In this paper, the authors tried to establish the hybrid method. To reach our goal, we studied some basic problems: (1) development of image measuring system, (2) development of computer software for calculating height distribution from digital image, (3) discussion on effects of experimental conditions on measurement accuracy. Finally, the hybrid method was applied to shape measurements of a lens and a coin.

(page 677)

Biaxial Laser Speckle Strain Gauge

ICHIROU YAMAGUCHI, TAMIKI TAKEMORI* and KOICHI KOBAYASHI**

The Institute of Physical and Chemical Research, 2–1 Hirosawa, Wakō-shi, Saitama 351–01, Japan
** Hamamatsu Photonics KK, Japan*
*** Tōyō Seiki Co.*

The laser speckle strain gauge, which measures surface strain in a noncontacting and automatic way by optoelectronically detecting and comparing speckle displacements at two symmetrical scattering angles, has been extended to a biaxial configuration. In-plane strains along the orthogonal directions have been measured by using four linear image sensors and high speed correlators with the sensitivity of a few tens of microstrains and a time response of a hundred Hz. The instrument has also been applied to simple and quick determination of Poisson's ratio of polymer films at various loading frequencies.

(page 682)

No. 9

An Experimental Study for Beam Displacements of Corner Reflection in Ultrasonic Angle Beam Testing (in Case of Variation of Test Specimen's Thickness)

MASAKAZU TAKAHASHI

Nippon Chutetsukan KK, Kuki Shobu Works, 1 Shōwanuma, Shōbu-machi, Minamisaitama-gun, Saitama 346–01, Japan

An experimental study of probe distance shift and beam path length variation by beam displacements of corner reflection was carried out with 10 to 50 mm thick specimens in ultrasonic angle beam testing. Even if the corner reflection specimen was 10 mm thick, the probe distance was shifted by beam displacement and then the beam path length varied. In total reflection condition of both surfaces, a right angled corner in steel specimen, the probe distancee shift and the variation of beam path length can be explained with Artman's theory for total reflection. In that condition of large shifts by beam displacement, we should take those shifts into consideration in evaluation of defect location and height in ultrasonic angle beam testing.

(page 724)

Effect of Stress Ratio on Fatigue Crack Growth in Polycarbonate. 1st Report. Special Characteristic on Fatigue Crack Extension Behavior with Stress Ratio

SHUNJI NAGASAKA

Kōgakuin University, 1–24, Nishishinjuku, Shinjuku-ku, Tōkyō 160, Japan

The effects of stress ratio on the fatigue crack growth in a polycarbonate specimen was analyzed. The stress ratio decreased the fatigue crack extension rate. The crack extension behavior was explained by the Paris formula with a effective stress intensity factor $U_p K_{max}$. The crack extension behavior was also explained by the index m 0.5 the Paris' formula, and was independent of the stress ratio.

(page 732)

No. 9A Proceeding of the fall Meeting.

No. 10

The Relation of Annealing Temperature and Fringe Order of Polycarbonate Used as Photoplastic Material

XIANG WAN and SUSUMU TAKAHASHI

Kantō Gakuin University, 4834 Rokuuramachi, Kanazawa-ku, Yokohama-shi, Kanagawa 236, Japan

Polycarbonate is a kind of glass clear, tough, ductile material. In recent years, it is increasingly used in photoelastic and photoplastic applications. Polycarbonate is available commercially in extruded sheet and stick form. However, this material contains a high degree of residual birefringence and must be annealed before use. But the examinations have shown that it is some difficult to clearly take off the residual birefringence, specially for thick polycarbonate sheet. The aim of this paper is to discuss the relation of annealing temperature and fringe order of polycarbonate obtained on market, to find the best annealing cycle for eliminating residual birefringence, and also to propose the best strain freezing cycle.

(page 863)

Studies on Improvement of Strain Sensitivity of Brittle Coating by Refrigeration Technique (Improving Mechanism)

KAZUO NAGASHIMA, AKIRA KANNŌ* and YASUHIRO INOUE*

Tōyō Unpanki Co., Ltd, 3, Ryūgasaki-shi, Ibaraki 301, Japan
** Kyōto Institute of Technology, Matsugasakigoshokaidōmachi, Sakyō-ku, Kyōto 606, Japan*

When brittle coating technique is applied to test surfaces showing lower strain than strain sensitivity of the coating (700 to 800 micro strains in general), it is necessary to improve the sensivity to obtain the crack patterns of the coating.

Refrigeration techniques employing water and/or air cooling are superior to physico-chemical method applying a solvent-cracking phenomenon in practical use. The authors have analyzed, therefore, the effects of the refrigeration through both the thermal stress of the coating caused by the temperature drop and the relaxation of the stress generated.

As the results of the studies, it was verified that the improvement of the sensitivity was dependent on the interrelation between the thermal stress developed on the coating and the relaxation of the stress under the process of refrigeration.

(page 899)

Numerical Simulation on Probe Index and Beam Angle of Ultrasonic Angle Probe

HIROAKI FUKUHARA

National Research Institute for Metals, 2-3-12 Nakameguro, Meguro-ku, Tōkyō 153, Japan

A numerical simulation method for the ultrasonic angle beam examination is proposed, by which sound field, directivity of angle probe and echo-dynamic curve of any reflector can be calculated. Application of the simulation method to the analysis of the determination process for the probe index and beam angle of angle probe which are usually decided by standard test block STB-Al is done successfully with suitable experiments, making clear that they are affected by many factors related with elastic wave propagation.

(page 875)

No. 11

A New Method for Absolute Calibration of Wide Band AE Sensors Utilizing Water Hammer Pressure

YASUHISA HAYASHI and MIKIO TAKEMOTO

Faculty of Science and Engineering, Aoyama Gakuin University, 6-16-1 Chitosedai, Setagaya-ku, Tōkyō 157, Japan

Water hammer pressure (WHP) generated by the collision of high velocity jet is well known to be an impulsive force with extremely short duration. This study aims to propose a new calibration method of wide-band AE sensors utilizing both water hammer pressure and deconvolution processing. Extensive experiments on WHP showed that the source wave of water hammer of small diameter jet was impulsive force like delta function. The source wave of WHP was also found to well agree with a theoretically predicted one, i.e. the maximum WHP and the duration time are determined by jet velocity and diameter of the jet respectively. Sensor calibration was done by the deconvolution of detected wave at seismic and epicentral surface by the WHP source wave and Green's function of the media.

Two types of wide band sensors, i.e. displacement sensor and multi-resonant sensor were calibrated by this method, and found to reveal their own frequency characteristics. The source wave of pencil lead breaking was also obtained by the deconvolution of detected wave by the transfer function which was determined by WHP, and found to well agree with the previous data. Proposed method was confirmed to be a reliable absolute calibration method of wide band AE sensors.

(page 921)

Fundamental Study on Detectability of Subsurface Defects (Report 1) Application of Magnetic Testing

SHIZUO MUKAE, MITSUAKI KATOH, KAZUMASA NISHIO and MASATO KOHNO

Kyūshū Institute of Technology, 1-1 Sensuimachi, Tobata-ku, Kitakyūshū-shi, Fukuoka 804, Japan

Calculations and experimental measurements were made on leakage flux density due to a subsurface defect. Specimens having a subsurface defect were made using diffusion bonding between a smooth plate and the other plate having slit like surface defect machined using an electrospark machine. Features of leakage flux due to a subsurface defect were made clear using FEM. Leakage flux decreases with the increase in distance, d_1, between the specimen surface and the top of a defect. When the height of a defect is the same, the maximum of horizontal component of leakage flux density at arbitrary lift-off normalized by that at the minimum lift-off experimentally obtained increased with the increase in d_1. The results of calculations for leakage flux density using FEM corresponded well with that measured using a Hall probe.

(page 930)

Fundamental Study on Detectability of Subsurface Defects (Report 2) Application of Ultrasonic Testing

YOSHIKAZU YOKONO, YASUO MINAMI, SHIZUO MUKAE*, MITSUAKI KATOH* and KAZUMASA NISHIO*

Non-destructive Inspection Co., Ltd., Kitakyūhōji-chō, Chuō-ku, Ōsaka 541, Japan

Ultrasonic testing is commonly used for detecting inner defects. Recently, new methods are tried to apply for detecting subsurface defects. In this paper, several ultrasonic testing methods were applied for detecting the subsurface defects in the testblocks, which were produced using diffusion bonding, and for estimating their size and locations. When using broadband type probe, edge of a defect located 1 mm or deeper can be detected by tip echo method. Surface wave method can be effectively used to determine the upper edge of a defect. Mode conversion method can be used to estimate the lower edge of a defect. Using creeping wave method, echo amplitude is approximately proportional to defect height in the case of small defects. Conventional angle beam method (frequency: 2 MHz, refraction angle: 70 deg.) is not effective for estimating defect size and location but available for detecting subsurface defects.

(page 937)

X-Ray Stress and Elastic Constants of Partially Stabilized Zirconia

MASANORI KURITA and HIROKI MATSUBARA*

Nagaoka University of Technology, 1603–1 Kamitomikoa-chō, Nagaoka-shi, Niigata 940–21, Japan
** Mitsubishi Electric Company, 2-2-3 Marunouchi, Chiyoda-ku, Tōkyō 100, Japan*

The X-ray stress and elastic constants of a partially stabilized zirconia (PSZ) containing 3 mol % yttria were determined for use in X-ray residual stress measurement. The peak positions of the diffraction lines were determined for the tetragonal (026), (133) and (044) planes with copper, chromium and cobalt K_x radiations by using the Gaussian curve method. The diffraction angles of the diffraction lines of these planes are 140.8°, 152.5° and 159.8°, respectively. The 95% confidence limits of the stress constant K for these planes were -608 ± 29, -295 ± 7 and -244 ± 4 MPa/deg. respectively. The relative errors $\sigma_K/|K|$ for these planes were 2.5, 1.2 and 1.0%, respectively, where σ_K is the standard deviation of K. To determine the stress, X-ray stress and elastic constants, it is important to choose a diffraction line having a diffraction angle 2θ as high as possible. The use of the tetragonal (044) plane with cobalt K_x radiation allows the most precise measurement of the stress by X-rays. There was the difference in the X-ray elastic constants for the three planes beyond their 95% confidence intervals of the constants.

(page 944)

Discrimination of Individual Photoelastic Fringes in Condensed Area Using Transformations of Coordinate System

YOSHIKA SUZUKI, SHIZUO MAWATARI, YOSHIAKI TOYODA and
MASAHISA TAKASHI

Aoyama Gakuin University, 6-16-1 Chitosedai, Setagaya-ku, Tōkyō 157, Japan

Not only in photoelastic analysis but also in many other optical measurements, the extremely condensed fringes are often observed and they are apt to hold important information for those experimental analyses. In order to facilitate the discrimination of individual isochromatic fringe particularly in condensed area, several types of transformation of coordinate system for the digital data of gray level are investigated. When utilizing a microcomputer for image processing, careful treatments on digitized data which are not continuous but discrete are necessary, regarding the one-to-one correspondence in direct and inverse transformations between the original and the transformed data. Also an additional caution must be paid on the fact that experimental data include unavoidable noises.

(page 952)

No. 12

Manhole Cover Deterioration Diagnosis Technique

KATSUHIKO HONJOH, YOSHIKAZU SUDOH* and JUN-ICHI MASUDA

*NTT Telecommunications Support Headquarters, Technical Assistance & Support Center,
3-9-11 Midori-chō, Musashino-shi, Tōkyō 180, Japan
* AIREC Engineering Co.*

This paper describes the technique of diagnosing the life-span of a manhole cover through the hammer striking method. Manhole covers installed on the road surface suffer damage such as cracking and enlargement due to the frequent overhead passing vehicles for a long period. Besides visual check, no other convenience has been developed which enables to measure quantitatively cracks in the cover. The speed of crack growth was theoretically determined by following an examination of the destruction process of cast iron. The examination showed the formula between stress intensity factor and stress cycles. The grade of the cover damage was quantitatively evaluated through the relationship between the resonant frequency and the attenuation constant by striking a cover with a hammer. On basis of the results through the cyclic load test and the field test, the manhole cover deterioration diagnosis technique was developed and a handy tester was realized which measures the damage grade of the cover in the field.

(page 986)

Simulation and Analysis of Acoustic Emission by Photoelastic Visualization Technique

KEIICHI SHIMOMURA and KAZUHIRO DATE*

*Department of Engineering Science, Faculty of Engineering, Tōhoku University
* Department of Mechanical Engineering, Miyagi National College of Technology,
48 Nodayama, Medeshimashote-aza, Natori-shi, Miyagi 981-12, Japan*

This paper describes the results of observation and evaluation of simulated AE waves by applying the photoelastic visualization method, developed for ultrasonic wave visualization and its quantitative evaluation. Pyrex-glass specimen was chosen as a visualization medium in this experiment and modeled on a compact tension specimen geometry as is similar to fracture toughness specimen. A piezoelectric ceramics, putting into the slit tip in the specimen, was driven by the electric pulse from an ultrasonic flaw detector to generate the simulated AE waves at the slit tip. Visualization was carried out using the optical system that consists of the linear polariscope with stroboscopic light source, in which flashing was synchronized with the simulated AE wave. Simulated AE waves were observed very clearly as a still picture, because the simulated AE waves were generated repeatedly and stroboscopically frozen. Burst type AE waves, similar to actual AE waves, were received at the AE transducer located at the specimen surface. AE event was observed as a complicated wave motion in the specimen, such as generation of longitudinal wave and shear wave from the AE source, the reflection and mode-conversion of these waves at the free surface of the specimen. The sound pressure distribution of the simulated AE wave was measured from the visualized waves using the photoelastic visualization method, and then compared with the waveform received from AE transducers.

(page 995)

Detection of Surface Flaws by Applying Magnetic Fluid

AKIRA KANNŌ and YASUHIRO INOUE

Kyōto Institute of Technology, Matsugasakigoshokaidōmachi, Sakyō-ku, Kyōto 606, Japan

As is well known, magnetic fluids are colloidal solutions with fine particles of ferrite in media and present both properties of ferromagnetism and fluidity. Few studies by applying the fluids have been made to detect the surface flaws as a non-destructive inspection, although they are generally employed to gas sealing, fine grinding, sorting and other engineering fields.

The authors have prepared various artificial flaws opened to the top surfaces on finite plates made of epoxide resin which is considered as a non-electroconductor (electric insulator), and have attempted to detect the flaws by scanning a probe coil of an eddy current testing device, after the magnetic fluids are infiltrated into them in the way of liquid penetrant test.

As a result of these studies, it has been verified that a newly developed method using the magnetic fluids can be utilized successfully to the non-destructive inspection for non-electroconductive materials.

(page 1001)

Study for Development of Low Profile, Portable Vehicle Weighing System

YŪ KANAGAMI and YOSHIROU NOJIRI*

Kyōwa Electronic Instruments Co., Ltd., 3–5–1 Chōfugaoka, Chōfu-shi, Tōkyō 182, Japan
** Dōro Keisō Co., Ltd.*

This vehicle weighing system is the instrument for measuring the total weight of a freight vehicle by adding each axle weight. The primary weigh detector was 80 mm thick and heavy, so it was too difficult to carry it. In the second system, the detector had six load cells, which measured weight by using shear strain type strain gauges. Since, the thickness of the detector was 30 mm, it was more strengthened against braking force. We installed them over the road surface and measured the total weight of 2 axle, 3 axle, and 4 axle freight vehicle. The accuracy was maximum 1.54% of 4 axles vehicle. To measure the excess weight of the freight vehicle overloading, this sytem has the indicator/recording section which calculates and prints out automatically.

(page 1009)

Mixed Mode Stress-intensity Factor (K_I, K_{II}) Measurements by the Strain Gage Consisting of 5 Grids to the Single Axis of the Chain. (1st Report. The Method by Measuring x, y Strain Components near the Crack Tip)

SHIGERU KUROSAKI, HIDEAKI NOZAKI and SHUICHI FUKUDA*

Tōkyō National College of Technology, 1220-2 Kunugida-chō, Hachiōji-shi, Tōkyō 193, Japan

** Welding Research Institute, Ōsaka University, 11-1 Mihogaoka, Ibaraki-shi, Ōsaka 567, Japan*

A technique for determining the values of the Mode I and II stress intensity factors (K_I and K_{II}) by using the electrical chain strain gages is proposed. The type of the strain gages used in this study is the chain strain gages consisting of 5 grids that have the same direction. The strains in the x and y direction are measured by the gages.

The method for the determination of stress intensity factors, K_I and K_{II} for the center inclined crack and the single edge inclined crack are examined.

The accuracies of K_I and K_{II} determined by the present experiments are within ± 10 percents and ± 20 percents, respectively, compared with the analytical values.

(page 1014)

Mixed Mode Stress-intensity Factor (K_I, K_{II}) Measurements by the Strain Gage Consisting of 5 Grids to the Single Axis of the Chain. (2nd Report. The Method by Measuring Same Strain Components near the Crack Tip)

SHIGERU KUROSAKI, HIDEAKI NOZAKI and SHUICHI FUKUDA*

Tōkyō National College of Technology, 1220-2 Kunugida-chō, Hachiōji-shi, Tōkyō 193, Japan

** Welding Research Institute, Ōsaka University, 11-1 Mihogoaka, Ibaraki-shi, Ōsaka 567, Japan*

The purpose of this paper is an experimental analysis of the mixed mode stress intensity factors by measuring a strain component. The chain strain gage consisting of 5 grids is used. The strain gages are bonded in the direction of the angle of $\pm \pi/2$ to the crack line. The mixed mode stress intensity factors, K_I and K_{II} are determined with the extrapolation method from strain values. Tensile tests with the center inclined cracked test specimens are carried out for examination of the accuracy.

The accuracies of the stress intensity factors, K_I and K_{II} determined by the experiments are within ± 30 percents, compared with the analytical values.

(page 1020)

INDEX